AF778371

MAN-EATERS AND MEMORIES

The man-eater of the Gori valley

MAN-EATERS AND MEMORIES

by

J. E. CARRINGTON TURNER

INDIAN FOREST SERVICE

ILLUSTRATED

NATRAJ PUBLISHERS

Dehra Dun

Text copyright © Natraj Publishers

First Published 1959
Reprinted 2007

ISBN: 978-81-8158-102-0

All rights reserved. This book is sold subject to the condition that it shall not, by way of trade or otherwise,be lent, resold, hired out, or otherwise circulated without the publisher's prior written consent in any form of binding or cover other than that in which it is published and without a similar condition including this condition being imposed on the subsequent purchaser and without limiting the rights under copyright reserved above, no part of this publication may be reproduced, stored in or introduced into a retrieval system, or transmitted in any form or by any means (electronic, mechanical, photocopying, recording or otherwise), without the prior written permission of the publisher of the book.

Published by Mrs. Veena Arora, for Natraj Publishers, Publications Division, Dehra Dun and printed at Chaman Enterprises, New Delhi.

DEDICATED

TO THE MEMORY OF THE VICTIMS OF THE MAN-EATERS AND BEARS MENTIONED IN THESE EPISODES

CONTENTS

LIST OF ILLUSTRATIONS

ACKNOWLEDGMENTS

Illustration No. 7 is reproduced by courtesy of the American Museum of Natural History, New York; No. 14 is reproduced from a photograph taken by Robert Bellairs

AUTHOR'S NOTE

FROM the very nature of his calling, involving, as it does, extensive travel in remote areas, a Forest Officer enjoys exceptional opportunities of meeting and studying an infinite variety of wild life. His daily routine in the forests builds up a unique store of experience, always interesting, often thrilling and sometimes dangerous. His knowledge of the forest fauna is constantly increasing, and inevitably he meets cases of tigers and leopards turned man-eaters.

With the advent of the ciné-camera hunter and the television audience, much has been done to bring to the great urban population a vivid portrayal of wild animal life; despite their excellence, however, these records cannot avoid some suggestion of a "forced effect".

The episodes which I here relate were all in the day's work to me, and I draw them out of a deep store of memories built up over a period of more than thirty years in the Indian Forest Service.

They are all true: and I only wish I could make the written word as vivid and thrilling as the experience.

Finally, I do not wish to give the impression that all tigers and leopards are savage and evilly-disposed towards human beings. They are noble creatures, beautiful, graceful and clever; in the ordinary course of their lives they feed only on their natural prey. But when Nature's balance is upset it throws up the maladjusted ones, and these were the creatures, menacing, swift and deadly, which brought fear and misery to the lonely hill villages, until they were tracked down and their lives ended by the even swifter and more deadly blow of the heavy bullet.

Happy Valley, Manderston,
Natal, South Africa.

CHAPTER I

THE BLACK MAN-EATER

"WE WILL PRAY for your long life if you will but shoot this animal."

The venerable spokesman of the village deputation looked earnestly into my eyes, and his companions solemnly nodded their assent. Against the background of their ancient temple the priests from the Máhábleshwar shrine stood studiously non-committal. They knew that I knew they were opposed to the taking of life in any form. We were good friends, and it was sound diplomacy on my part to remain so, as these holy men wielded considerable influence over the villagers.

"I have been greatly pleased," I replied, with careful emphasis and my gaze fixed firmly on the middle distance, "to note that illicit fellings in the Government forest have much decreased, and it is to be hoped that the time is not far distant when such irregularities will cease entirely."

This apparent irrelevance was greeted with the silence it deserved. Encouraged by discreet nudging from the back row, and only slightly abashed, the spokesman prepared to resume his plea.

"We are afraid to enter the forest to collect the dry fallen wood for our requirements."

From beside me there came a sudden growl as my attendant Forest Guard leapt to his feet.

"It is a lie. Daily I see you in the forest making the fallen wood into headlands of great size, and returning laden to the village."

"The tiger is exceedingly fierce——"

"The animal is harmless," burst in the Guard. "Several times I have met him, and I know him to be blameless. He kills deer and wild pig but nothing else. The forest holds sufficient animals for his needs."

I glanced towards the priests and asked them whether this tiger had any reputation as a man-eater.

"No," they replied, and there was general though reluctant assent from the elders.

"In that case I can see no reason for killing him. The animal is following the natural pattern of his life, hunting his prey in the forest, and so reducing the damage done to your crops by deer and wild pig. Such an animal must surely be regarded as a protector of your livelihood."

"And a protector of the forest," put in the Guard, fiercely surveying the deputation.

Accepting their defeat with good grace, the elderly villagers retired with protestations of undiminished respect and friendship, and the priests went back to their temple from which soon came the delicate notes of their devotional bell-ringing.

On the seven-mile trek back to my headquarters at Máhábleshwar, the Guard enlarged, though more mildly, on the subject of the tiger and the villagers. The tract of forest concerned was of a peculiarly elfin type. The trees, being on the extremity of the plateau, had to bear the full force of the strong monsoon winds, so, instead of growing upright as jaman trees normally do, they developed a stunted and very slanting posture, a symbol of adaptation to the local climatic conditions. The tiger had a definite liking for these sloping stems. They were large enough in girth for him to climb effortlessly on them. He would then scratch deeply into the soft bark to manicure his claws. Sharpening them thus, and incidentally ridding them of his latest prey's putrid flesh remaining in the concavities, he would sit down lengthwise along the stems in perfect comfort to survey his surroundings like the magnificent monarch he was.

It was clear that the Guard, whose responsibility this forest was, knew the tiger intimately. He described his habits to me and pointed out many trees with deep, long, parallel incisions made by the tiger's claws. His pug-marks (footprints) told me he was a big male, so very worthy of being styled "protector of the forest". The villagers named the jaman tract "the tiger's jungle".

The tiger continued to make his home in the forest and there continued to be a decrease in the number of illicit fellings, for which improvement in conservation I silently expressed my gratitude to the highly efficient but unpaid guardian of a very exceptional forest.

But that was before the Terror came, stalking silently out of the forest, bringing sudden and horrible death to the villages in the valley.

The Koyna valley is one of the loveliest I have seen. It is situated to the south of the extensive and attractive plateau of Máhábleshwar which overlooks it. The entire landscape is marked by a pleasing configuration, made more delightful to the eye by the dense, dwarfish forests clothing the slopes like a green and gorgeous mantle.

The valley forms the entire catchment area of the river Koyna, from its very source to a point several miles southwards where its waters merge with those of a much larger river, the Kistna, to flow in the direction of Miraj, a large and growing town noted far and wide for the excellent medical work carried on by a devoted band of missionaries.

Máhábleshwar, where I had my headquarters, stands at about 4,500 feet above sea level on a plateau from which on a clear day the Arabian Sea is visible. The plateau occupies the summit of a ridge on the western mountain range known as the Western Gháts. The climate is excellent all the year round except during the months of June to September when the rainfall reaches the almost incredible figure of 300 inches, due to the plateau being exposed to the full force of the south-west monsoon. The mean annual temperature is 67 degrees Fahrenheit, and the recorded maximum is 90.

In spite of the exceptional rainfall, a deluge of fourteen inches a day being a common occurrence, the plateau is remarkably well drained. Very little of the red laterite soil is washed away. This efficient soil-conservation is entirely due to the presence of its forests with their complete canopy and deep sponge-like layer of humus and leaf litter. They have been most carefully managed by the Forest Service. Though the entire country is hilly yet landslips are few and far between.

For three years I was the only European to spend the whole twelve months in Máhábleshwar. It was an interesting and unique experience, with a pleasing pattern, developed over the years, of alternating indoor and outdoor life. My duty-tours of the forest in April and May were entirely delightful, while during the monsoon period I employed the long indoor hours most fruitfully in mastering the Marathi language.

But the pattern was not wholly smooth.

At the opening of the second year I began to hear disquieting tales from the valleys. Disease, never at a safe distance, had struck the cattle, and communicated itself to the forest. My Guards reported a great diminution in the numbers of deer, and even the wild pig appeared to have been affected. Then the Terror struck. In a remote hamlet at the far end of the valley a child was snatched from the fields. A week later a similar incident was reported from the adjoining territory. Isolated tragedies such as these were not, alas, unusual, but within a week they were no longer singular. Another hamlet in the Koyna valley was twice stricken, and the thing began to assume different proportions. The news filtered up from the villages, and it required no great imagination to hear the wailing of the parents as they beat their breasts in their grief.

Official attention was directed towards the problem, and various excursions were made by summer visitors who were keen to shoot a man-eater. But the terrain was against them, and he survived all their efforts, traversing the length of the valley and inexorably adding to his victims. By April the known number had risen to no less than sixty-three; the man-eater became a daily topic of conversation, assuming in all our minds an urgent priority.

Constantly I pondered on a plan that would bring me within trigger-distance of the raider, but even on such a matter I was not entirely my own master and there were many tasks to be fitted in before the blanket of the rainy season descended.

But the day came when I found it necessary to visit the forests in the valley and I set off with every intention of succeeding where others had failed.

There were no roads leading to the Koyna valley. Access was possible only by paths not more than two feet wide, and these followed severe gradients. There were the faintest signs of a very old road running eastwards, but this, after two and a half centuries, was nothing more than slippery scree. This ancient track is of historical interest. Its alignment showed no science. It merely followed a ridge or spur, wherever the ridge or spur happened to go. But it bore the name of the Great Siváji, the all-conquering Máratha chieftain, who made and used it in the seventeenth century. Swiftness was his motto in all his wars, and so his lines of communications invariably followed where physical resistance was least. He had to move thousands of soldiers on foot from battle site to battle site. His soldiers' monkey-like ability to climb cliffs specially fitted them rapidly to surmount this age-old scree.

Often have I traversed it in company of my Forest Guards. Not one of them will walk along this track with you unless he stresses the great veneration in which Siváji is still held. For are not all Máráthas of this land the direct descendants of Siváji's scions? They of the present day can, I vouch, also negotiate cliffs in a simian-like manner. Always martially-minded, they will relate stirring tales of Siváji's deeds. They will tell you that the whole of the Koÿna valley, the plateau of Máhábleshwar and all the coastal belt known as the Konkan, formed part of Siváji's kingdom.

Along this track I plunged and scrambled with my trusted Forest Guard, Kusaráo.

On the way to our second camp we were halted by a distraught messenger from a village some four miles distant. It was enough to question him on the salient points of his message. In his unhappy face I saw afresh the tragedy of cruel bereavement. A twelve-year-old girl had been snatched and dragged into the forest.

I changed direction immediately and we were soon at the outskirts of the village. I could get little sense from the few villagers prepared to talk. It was as though they had all been stricken. The men had made no attempt to pursue the beast or to retrieve whatever might remain of the corpse. When, however, I showed that this was my intention they seemed eager to follow me, giving me vague and contradictory indications of the direction taken, but I told them that I would take only Kusaráo, who was carrying my second gun.

Kusaráo was a typical Márátha, short and sturdy in stature, absolutely staunch in a crisis. I knew I could rely on him to stand fast and hand me the second gun if necessity arose.

It is a wise plan to have, if possible, a second weapon in reserve when tracking a man-eater. You owe the maximum degree of safety to yourself, and the greatest measure of success to the suffering villagers. After all, a gun is only a piece of machinery, and, however well tended, liable to develop a fatal defect at any moment. But it is essential that the carrier is reliable, and not likely to desert you in a crisis.

I proceeded in the general direction indicated and halted on the edge of the undergrowth, hoping to be guided by the stirring of a bush, since the killer would be changing his position from time to time while devouring his prey. But I could see no movement. I then signed to Kusaráo to tell him I was entering the forest in the hope of actually seeing the man-eater.

Step by step we pressed through the scrub, zig-zagging to enlarge our view, concealing ourselves behind the stems of the trees and listening intently. There was neither a movement nor a sound anywhere. There was nothing for it but to advance further into the undergrowth and zigzag more widely. We had not gone twenty paces when we came upon the girl's body. The sight before our eyes was horrible in the extreme. The Terror had obviously gorged himself and gone to a secluded spot to lick himself clean, and sleep. I decided that he would not be very far away after such successful hunting, and so planned to search for him uphill.

Even more slowly and cautiously we proceeded, alert for any eventuality, and then, after patiently searching for nearly half an hour, I saw a slight movement among the tops of the bushes to my left front. I immediately increased my pace in that direction, but here the undergrowth was so dense that all I saw of him was a flash when he rushed along the contour. Then he was out of sight. I had my gun raised to my shoulder but there was insufficient time to aim. It all happened in a second. The tall bushes were in his favour. To fire haphazardly in such circumstances at carnivora, especially at man-eaters, is nothing short of crass stupidity. Had I fired I would merely have scared him, would not have hit him, and would certainly have made my future chances more difficult. Looking in the direction in which the killer had bounded out of sight, I said in Máráthi: "We will meet again." Kusaráo smilingly concurred, adding tactfully that he had seen the bushes disturbed but had not caught a glimpse of the man-eater.

By this time it was late in the afternoon. The setting sun was close to the horizon and the light under the forest canopy was fading. So we retraced our steps down to the village. The men had removed the terribly mangled and scanty remains of the poor girl.

Tying up a goat to a stout peg, driven into the ground on the edge of the forest, brought no success. I had the goat in full view till darkness fell, and then returned it to its owner. In the circumstances the Terror was unlikely to react to such a bait.

It was time for me to return to my camp which was about five miles away. The villagers very considerately supplied me with a lantern and, accompanied by Kusaráo, I forded the Koyna river and reached my tent in an hour and a half. On the way I deliberately turned my thoughts away from the failure of my first attempt and concentrated on the best way of meeting and destroying this infamous man-eater. There was an interesting feature about him that set me a-thinking. I had seen but little of him; just a very small portion of his back when the bushes had slightly parted, but I had noticed that he did not appear to be the usual colour of his kind; I could not, however, be quite certain about this particular point.

Entering my tent, I observed that the daily post had dutifully arrived on the table. Among the envelopes there was one that bore an important official crest, and so demanded priority. It contained a letter from Robertson, Indian Civil Service, Private Secretary to the Governor, informing me that His Excellency desired me to accompany him through the forests in the immediate vicinity of Government House to explain the silvicultural treatment being applied to them. On a previous occasion, at a tea party at Government House, His Excellency (Sir George Sydenham Clarke, afterwards Baron Sydenham of Combe) had evinced a keen interest in forestry. The date and time mentioned by Robertson were so fixed that I had no option but to pack up camp early next morning and return to headquarters by the shortest route to obey the summons. Such a letter, however nicely worded, from a Governor's private secretary to an official, is very definitely to be interpreted as an order. So, to my profound chagrin, the man-eater gained a further respite.

It was mid-May before I was able to return to the Koyna valley after completing certain overdue inspections of forests on the plateau. Reports received from my Forest Guards told me that the man-eater was active in my territory again, but he was killing at considerable distances apart. I concluded that my best plan, and one that would give me the utmost mobility, would be to camp about the middle of the valley and see what happened.

I did not have long to wait.

On the fourth day news came by runner that the Terror had killed at a village seven miles away. Hastily packing some sandwiches and a generous supply of biscuits in my haversack, I set out for the scene of the tragedy. Kusaráo carried one of my guns, and another Guard, Rámji, shouldered the other. In the circumstances we forced the pace to the utmost. Fortunately we were all in fine fettle and found no ascent an obstacle to our rapid progress.

On arrival at the village we were surprised to find it empty except for wailing women, small boys and girls. I immediately questioned the boys. Stuttering from fear, one of the older boys informed us that Báji's son, aged ten, had been taken by the man-eater, adding that all the men, accompanied by youths, had gone into the forest to recover the body. This, as we suspected, explained the noises we had heard as we approached.

With an expression of utter disgust on his face, Kusaráo said: "These people are very foolish. Why didn't they wait for us before they entered the forest and made such a hullabaloo? We could have quietly approached the animal while he was busy eating the body. What a good chance that would have been to kill him." I agreed with him and expressed my disappointment, made all the greater after our seven-mile march at considerably more than normal speed.

There was nothing to be done but to proceed to the spot where the body lay. We found poor Báji weeping his very soul out. His friends and some boys were busy cutting saplings and portions of climbers to make a bier. I spoke comfortingly to Báji. The dead boy had been badly mauled and the flimsy loin cloth was torn to shreds. Báji was much too upset for me to pass on Kusaráo's whispered suggestion that, as the killer had eaten very little and was sure to return, he should allow the body to remain where it was and I would wait for the man-eater to return. I tentatively made the suggestion to one of Báji's friends but was not surprised to be met with a flat refusal. The boy was a great favourite in the village and, added the friend: "We could not think of leaving the boy here; we must take him to the village, and we have already sent for a Brahman priest from the neighbouring village to come at once and perform the last ceremony."

I beckoned to Kusaráo to follow me. We searched the forest for over an hour, hoping to meet the killer, but we drew a complete blank. The din made by the shouting and yelling villagers, with violent beating of drums, as they proceeded through the forest to find the corpse, had thoroughly scared the Terror. The men had actually gone well beyond the dead boy, making the noise of their lives, to drive the man-eater away over the ridge into the adjoining district.

We decided that as further search was in vain we should return to the village. We sat down in the courtyard. I found that I was famished, so I ate sandwiches and biscuits, and drank from my water-bottle. The villagers gave Kusaráo and Rámji a generous helping of milk from their scanty supply. After this we sat quietly in the shade and I asked for details of the tragedy, the first to occur in this village for a considerable time.

The men informed us that they were, as usual, busy in their fields, hoeing and preparing them for the coming monsoon crops. The women were cooking their midday meals in their houses. The boys were playing, chasing one another round the huts, and, occasionally, in their excitement, extending their runs behind the houses. It was during one of these pursuits that the man-eater, lying cleverly concealed on the edge of the forest and awaiting his opportunity, sprang on the boy, and carried him away into the undergrowth. I inquired if anyone had seen the killer. A boy, who had been chasing Báji's son, had seen him, but could only say he was dark-coloured with a long tail. I remembered my own fleeting glimpse of a darker-than-usual animal.

I tarried a week longer in the valley, tying up goats as bait in various places, especially in Báji's portion of the forest, but not one of them was attacked. I waited for news of a further tragedy, but none came. It was clear that the Terror had left the valley to carry on his depredations elsewhere. As a rule man-eaters are great travellers, and this one was no exception. My second attempt had ended in failure.

It was now the fourth week in May. I had inspected a large tract of forests in the valley, and it was time to return to headquarters before the onslaught of the monsoon rains. The first of the clouds had begun to appear, reliable precursors of heavy downpours. It would be another week before I could reach my destination.

On my return to Máhábleshwar I had a long discussion with the Chief District Magistrate, S. R. Arthur, of the Indian Civil Service. He was directly responsible to the Government for the peace and prosperity of the entire district of Satára of which Máhábleshwar was a subdivision. No head of a district had greater solicitude for the welfare of the rural population, or deeper sympathy for them in their sufferings.

I described my two abortive attempts, and we again went over the possibilities of bringing about a successful encounter. The thing had now got into my blood and I did not feel it rash to give him my promise that, as soon as the rains cleared and October came, I would go down again to the Koyna valley and do my utmost to put an end to the Terror.

About mid-June the Governor, his staff, the all-important Secretaries and their considerable office establishments, visitors from Poona and Bombay, together with the more well-to-do shopkeepers and hoteliers, all left Máhábleshwar. The exodus was rapid and complete.

Soon afterwards the first rain-laden clouds came over the plateau and precipitated their contents in, literally, sheets of water. So heavy were these downpours that the doors and windows of houses had to be protected by grass screens during the rainy season.

Though confined mostly to indoors I did not find time hanging heavily. I busied myself devoting long hours to translating the *Arabian Nights* into Maráthi, the local language in which I had already attained both proficiency and fluency. Often, in my assiduity, I found the days much too short, but despite this pleasant preoccupation I could not put from my mind the dark killer in the valley.

Then came October with its sapphire blue skies. In the meantime much outdoor work had accumulated and its disposal was urgent. There were many things to be done in the forests on the plateau before I could return to the Koyna.

My first excursion saw me escape death by inches. Accompanied by Guna Máhádeo, a deeply wrinkled and experienced Guard whose beat comprised the forests in the immediate vicinity of headquarters, I proceeded in a south-westerly direction and entered the forest to enable him to show me a score or so of trees suitable for a free grant to be given to a village called Awkáli. The villagers required the trees without delay to repair their rain-battered huts before winter arrived.

Guna Máhádeo indicated several trees and as these met with my approval I marked each of them with my special hammer on the bark near the ground. The impression thus made acted as a control over the villagers who were allowed to fell only the trees so marked.

Guna Máhádeo and I walked up and down the hillside looking for a few more trees to complete the free grant. He pointed out one, some twenty yards away. We examined it, and I found it quite suitable. I got into position, raised the hammer, and commenced the downward blow to the bark when I felt my striking arm pulled back with a violent jerk. "Look," exclaimed the Guard, pointing to the base of the tree, just below where the hammer would have struck. My eyes focused more carefully this time, and then I realized that it had indeed been only a matter of inches. There lay a deadly phoorsa (*Echis carinata*), a reptile of the viper family, coiled with its head slightly raised and ready to strike. As it lay motionless its light brown colour, with pale spots, blended perfectly with the forest floor. Considerably shaken, I quickly took my walking-stick from the Guard and killed it. These snakes, known as the saw-scaled viper, vary in length from fourteen to twenty inches, and taken weight for weight, their venom has been proved to be more virulent than that of the much larger cobra. In the absence of an antidote their bite is fatal. It is characterized by bleeding gums. I remember that several villagers died yearly from the bites of phoorsas. Undoubtedly I owe my life to the keen eyes and quick reaction of Guna Máhádeo.

By the second week of October I had completed all urgent work on the plateau. My impatience had been mounting, for my next duty was to contact the man-eater. So my first camp was on the way to the valley. On arrival at the village site where my tents had been erected I found several elders of the village waiting for me. After the usual greetings they begged me to shoot a pig for them; adding, what I felt to be unhappily true, that it was a long time since they had tasted one. They volunteered to supply all the beaters, and undertook to dislodge the pig in the forest within an hour. With deep earnestness the headman of the village said: "We know where the pig is lying up, we must get him before he is taken from us." I replied I was quite ready to accompany them and do my best to shoot the animal, if they would beat the forest intelligently.

The usual consultations took place. It was decided that the best position for me would be half-way down the hill, concealing myself near the base of a large tree on the edge of a narrow stone-strewn ravine. The beaters would go silently in single file to the crest of the hill, taking care to make a wide detour in order not to disturb the slumbering pig. At the crest they would fan out and then beat downwards towards me. They would whistle to let me know the beat had begun.

I took up my allotted position as silently as possible. There was much whistling accompanied by high-pitched shouting and the beating of empty cans. I was alerted. It was not long before I heard the faint dislodging of stones in the ravine. Before a minute passed, a real monster of a boar came in sight but, to my dismay, he was coming down the hill straight at me. He was a poor target for me but I was a mighty good one for him. Crashing through the bushes his mighty bulk catapulted itself down the decreasing space between us, then, with a suddenness sufficient to overturn it, he met an intervening boulder and turned precipitately aside. As he did so he presented the broadside target for which I had hoped. Making no mistake I fired and killed him.

Yelling louder than ever, down sprinted the eager beaters to my position. I pointed to the animal lying in the ravine. I did not join in the yelling but I was not less pleased than they.

When they saw the size of the boar great was their joy and such their excitement that several youthful beaters indulged in a dance of impatient anticipation, nodding their heads repeatedly and looking at the boar. But the headman intervened, stopped their frolics and told them there was work to be done. Turning towards me for permission to cut saplings, he gave the necessary orders, and very soon they had been cut, and ropes made out of Acacia climbers.

Within half an hour all was ready to carry the great pig down to my camp. It took eight men to accomplish this with much shouting and cheerful groaning, each warning the others not to stumble. At last the monster was deposited about fifteen yards in front of my tent. By this time there was a large gathering which doubled and trebled in an amazingly short time. News had spread to the nearby villages, and men, women and children flocked to see the mighty boar from which the poles and ropes had been removed. There was much loud talking and gesticulating. It was a gathering of all ages but I was intrigued to see a goodly proportion of elderly men in the crowd. Totteringly they sat down, some of them with asthmatic chests and breathing heavily.

The problem of the just distribution of the carcass amongst the happy and hungry congregation at first baffled me. But I quickly decided to call the headmen of the villages to my tent, where we held a council of equity. I informed them that I wanted the tushes and nothing else. The headman soon decided that distribution was to be effected firstly on the basis of villages and, thereafter, amongst the individual families of a village.

There was, however, one proviso which completely perplexed me. Innocently, I asked for elucidation. They looked at one another. Then one of them, turning to the others, inquired: "Would it be immodest to explain?" At this the most senior of them spoke, smiling broadly at me: "Our humble request is that all the fat on the boar should be distributed only amongst the old men present. We Marathas, you understand, have a great veneration for old age, and it is a well-known fact that when old men eat the fat of a wild boar, especially such a one as this, which weighs about 400 pounds, they feel rejuvenated in every way."

But sterner work lay ahead. Although he had not yet been identified I knew that I was up against the cleverest of all the forest animals, the leopard, whose dark cunning had for so long held the valley to ransom at a time when the disease among their cattle was causing the villagers much distress.

I could not hope to be able to forestall the new attack, but I vowed to do everything that speed and experience could to make this third attempt the lucky one.

As I went through the straggling villages between my camps, I was aware of the tension caused by this ever-present threat. I saw the children afresh, hating to see one unattended; I saw the tiny temples adorned with flowers and coloured rags and papers. I knew that to these simple folk this was the visitation of an evil spirit, to be countered with daily prayers:

"O Shiva! O Vishnu! We pray you to exorcise for ever from our midst the destroyer of our peace and happiness."

The little bells tinkled disconsolately.

My guards had been well posted over the district and from their reports I felt certain I was in the man-eater's latest field of activity.

Kusaráo and Rámji went about the camp ready for immediate action. My guns were at my hand, and I had the wonderfully relaxed feeling that comes from the matching of physical fitness with a set purpose.

Tragedy came soon enough. From the very next village on our route, a bare three miles away, came the urgent news that the leopard had just killed a young girl, dragging her into the nearby forest.

My Guards were ready immediately and we set off at a brisk pace. It was only nine o'clock and the day stretched before us, an inestimable advantage for which I had devoutly hoped.

I had already sent earnest requests to the headmen of villages not to resort to shouting and drum-beating, nor to approach the corpse, but to await my arrival. I emphasized that I quite understood their feelings, particularly their deep grief at the loss of a beloved one, and their anxiety to recover the body before more of it was devoured by the man-eater. I added that it was quite impossible for me to see the Terror if they failed to obey my instructions. I did my utmost to explain to them that it was imperative to match the killer's stealth and cunning with an equal measure of skill and silence; and that the foe we were endeavouring to outwit was the cleverest of all the animals in the forest, even more wily than the tiger.

Very fortunately, the headman concerned had succeeded in persuading the villagers to heed these instructions. In the village itself the silence of the grave prevailed. The men were standing in a group in front of the headman's hut, while the women and children were indoors.

Quietly I asked for the fullest information. One of the men, more intelligent than his comrades, led us to the corner of the extreme hut and, peering nervously round it, pointed to a piece of cloth caught up in a bush to our right. This was indeed a valuable clue as it showed, with certainty, the direction in which the man-eater had taken his victim.

Never was time more precious. The killer must still be on the body, since he had dragged it into the forest but a short time ago. There had been no noise, and the silence was just what he desired for his meal. The first step in this battle of wits was to approach silently the piece of cloth on the bush, to stop there and listen intently for the slightest sound, and look for the least movement in the bushes. Kusaráo was by my side straining eyes and ears. Minutes seemed hours. There we stood waiting.

I had an almost overwhelming conviction that this was the only opportunity I would ever get to destroy the Terror. I was looking to the right and Kusaráo to the left. Patience I knew was essential. To enter the undergrowth ignorantly, and walk in any direction, would be fatal to my chance. We continued to survey the tops of the bushes. There was not a sound, not a movement to show us the man-eater's position. But the sun was still high in the heavens and time was on my side. Patience and more patience. There we continued to stand, statue-like, timeless in immobility, hoping that the fortunes of hunting this Terror would turn in our favour, and the prayers of the priests be answered. I tried desperately not to be mesmerized by the leaves and the light. As I continued to stare at the tops of the bushes, I felt my left arm gently nudged. I looked at Kusaráo. He cocked his eyes to a bush about thirty yards in front of him. I saw the movement and blessed his vigilance and keen sight. Our patience had been rewarded. The Terror was busy devouring; in changing his position he had caused the bush to stir. We now knew exactly where he was.

I quickly made my decision, and signed to Kusaráo that I was going straight in the direction of the bush. He nodded his willingness. Now the undergrowth was a serious problem. In moving through it, however cautiously, there would still be the unavoidable swish, swish, as we advanced. I decided we had no choice but to proceed as silently as we could. When I took a step forward Kusaráo did the same behind me. Thus we progressed, step by step, pausing after every two or three paces to listen. My eyes were glued to the one spot which meant so much. My rifle was half up to my shoulder, ready for any emergency. Slowly we got closer and still closer.

Uppermost in my mind was the constant anxiety that the killer's keen sense of hearing would defeat me, and that he would get away too far ahead. We had covered about half the distance towards him when he heard us. With a growl he moved suddenly away through the bushes. My hopes rose. The opportunity to shoot would come where the undergrowth was not so dense. I immediately quickened my pace to keep him within killing distance. It was now an easy matter to follow him. As he progressed, the tops of the bushes swayed gently, and so told us exactly where he was going. During these speeding seconds no two brains were more active. His, no doubt in wondering why he was being followed with only the swish, swish of the bushes and no noise, a situation he had not known before, In such novel circumstances what should he do next? Mine, bent on maintaining the same distance between us, trying to anticipate his next move. My eyes were now fixed ahead of him to see if there were any patches of ground where the undergrowth would thin out. I saw some bamboo clumps ahead and my hopes soared again as I knew the undergrowth there would be relatively sparse. These clumps were in his direct path, and he continued straight towards them.

Eventually he reached the bamboo and, at the same moment, his innate curiosity got the better of him. He was familiar with the shouting and yelling of villagers and the beating of their drums, but now there was no noisy performance. Instead, just two men persistently and silently following him. What did it all mean? He decided to have a look at us.

He stood, only partly hidden behind a bamboo clump, and turned his head towards us. I could only see his head and left shoulder through the culms of the clump. Here was my chance, perhaps never to return. I raised my rifle, aimed at his shoulder, as this was the better target, and fired. Down he went but tried desperately to rise and, in so doing, collapsed again. He was struggling hard to regain his feet, but the split-nosed bullet had done its duty efficiently. He continued to struggle. I wanted him in full view, so I hastened uphill to avoid a tree and then turned sharply towards him. He now presented a full target, still trying to rise, but in vain. With a second bullet I ended his agony.

Kusarão, ever steadfast, kept only a yard behind me in all my movements, ready to give me my second rifle. In his joy he uttered the Márâthi equivalent of "God be praised, the enemy is dead". We approached the lifeless leopard and found him to be jet black all over. There was not a semblance of a yellow hair in his coat. Later on, when we laid him down in full sunlight the black rosettes, so very clear on all ordinary leopards, were faintly discernible on him.

We then examined the Terror in detail. He was a fine specimen, powerfully built and middle-aged. From nose to tip of tail he measured seven feet, seven and a half inches. We gave him a first-class certificate of fitness. He was sound in every limb, and there was not a chipped or missing tooth in his mouth or a damaged claw in his paws. These parts, as well as his forearms and portions of his chest, were thickly smeared with the luckless girl's blood. His black colour was due to unusual pigmentation in the embryonic stage, a case of melanism.

Shrill calls up and down and across the valley quickly spread the news of the killer's death. Ere long, a crowd of villagers had gathered where the leopard lay. But it was a hushed assembly, as a few paces away were the remains of the dead girl. Soon these were carried from the forest and placed before the door of her parents' house and covered with a cloth. Friends had already arrived, and had begun preparing the last rites according to their religion. Advancing towards the grieving father, I expressed my sympathy with him and his family in their tragic bereavement.

Accompanied by the valiant Kusaráo and the youthful Rámji, I returned to my camp. As we reached it the sun dipped appropriately over the distant escarpment of the plateau. My promise to Arthur had been fulfilled. Peace and tranquillity had returned to the lovely, but sorrow-stricken, valley of the Koyna.

CHAPTER II

AN EVENING'S VIGIL

BETWEEN THE GODAVRI and Kistna rivers in central India, where the annual rainfall ranges between twenty and thirty inches, considerable tracts are covered by deciduous forests approaching in general character an arid type of vegetation. The trees, never very tall, are interspersed with a variety of hardy shrubs whose environmental adaptations include an intensive development of thorns to protect themselves against injury by man and animals.

During the hottest month of May and part of June, the flow of the main rivers is almost imperceptible, while their numerous tributaries dry to such an extent that the only available water for miles in every direction is a pool, or waterhole, in a dry river-bed, serving as a natural drinking trough to which all beasts and birds of the forest are driven by compelling thirst.

They visit this pool almost daily until with the arrival of the south-west monsoon rains all the tributaries flow again.

In such a region, May and June are critical months for all wild life, critical not only by reason of thirst but also by the ever-present danger of carnivora which resorted to the water-holes to stalk and kill their prey.

It was the third week in May, when with a friend employed in one of the big Kanpur textile mills, we decided to make an evening excursion from our camp to observe wild life. Leaving our guns behind, and dressed in khaki shirts and shorts, the coolest attire in summer, we walked along the railway in territory we knew well. Reaching a long bridge we took up our perch on one of its abutments, immediately below the track. These projections in the abutments, presumably designed to facilitate inspections by engineers, made excellent observation posts. With the aid of powerful binoculars all the birds and animals approaching the pool could be seen and studied closely. Our perch, about fifty feet above the dry bed of the river, gave us maximum concealment.

We were in our positions between half-past four and five o'clock. The waterhole was about forty yards downstream. The river-bed was otherwise bone dry as far as the eye could see. The banks on both sides were not more than four feet high and sloped down gently. Here and there trees lined both banks, their roots being partly exposed by erosion caused by periodical spates and floods. Some trees, no longer able to support themselves, had fallen into the river-bed. Filling the spaces between the trees, near and on the banks, were acres and acres of thorny scrub. Through this there were game tracks to the river. Raising our binoculars to look further afield we saw groups of trees, and beyond them low hills crowned by large rocks.

Watching wild life is an intensely interesting pastime and the secret of success is undoubtedly to be completely hidden, and remain absolutely silent. The watcher must train himself to remain perfectly motionless. The animals and birds will then move about with the naturalness desired. If watching cannot be done without smoking the chances of success will be greatly reduced, if not altogether nullified. A good motto to try and work to is: "Only wink and no more." Admittedly spartan in advice, but in truth essential, and in achievement not impossible.

We were particularly keen to study the sequence in which the many species of birds and animals arrived at the pool. Our vigil had not lasted a quarter of an hour when down swept a flight of blue-rock pigeons and alighted near the pool's edge to which they quickly advanced, dipped their soft bills into the water and then rose immediately to fly away in the direction of a grove of trees in the distance, to roost safely for the night.

There was then a pause, but not a silent one. The trees along the banks were full of twitterings from a variety of birds hopping excitedly among the branches and looking down. The reason for this agitation was soon apparent. A wild cat came slowly from under the bushes, walked leisurely to the water and, stooping, took his time lapping. The birds in the trees evidently saw him clearly as he was moving through the scrub. Having quenched his thirst he sat for a moment near the pool, scratched his neck with his hind paw, then walked away and was lost to sight.

As he disappeared to a safe distance the lively chatter in the trees almost ceased. Several bulbuls flew down, accompanied by tits, white-eyes and tailor-birds. They descended accurately on the water's edge. There was no need to walk even a fraction of a foot. They lowered their heads, and to them a drink was just a matter of two or three seconds and no more. It was dangerous to tarry by the water. They were safer where they came from, amongst the boughs. Their eyes were on the alert for their enemies, the cats and the sparrow-hawks.

The next thirsty ones to appear were a pair of hoopoes. They came through the air with a succession of characteristic swoops with wings closed. They alighted, then raised and lowered their attractive hoods, like a hand-fan opening and closing; satisfying their thirst, they flew off to a neighbouring tree. The king-crow, or black drongo, came next, a pair of them. For a minute or so they rested on a branch, then dived low over the pool to catch insects. Repeatedly they returned to the branch, rested, poised and dived again. Having fed well they then thought it time to drink. Quickly down and quickly up, they flew away, no doubt satisfied with life in their happy union and in their role as protectors of little birds against the predacious types such as the tree-pie.

Some mynahs and babblers (seven sisters) arrived almost together. They seemed in urgent need of a drink. The former had come from a longer distance, the latter from close by with a series of short flights, gliding and flapping alternately from tree to tree, and never failing to keep up a continuous chatter amongst themselves. I longed to understand their endless conversation. Shortly after their departure, a covey of jungle fowl came on the scene. They were taking no risks in the bushes, and so came to a halt in a tree. Perching longer than we expected, obviously to inspect the surroundings, they hopped down, quenched their thirsts, flew back to the same tree, and then made their next flight to another tree about thirty yards away. The covey consisted of a handsome self-conscious cock, and four obedient hens. The next in the sequence of visitors were a pair of doves, followed swiftly by another pair. All four approached at such speed that they over-flew the pool, and were obliged to circle once over it before they came to rest near the water. Their drink was the quickest of all. It was just a dip of the bill. Flapping noisily in a vertical ascent from the sand, they came straight over our heads; it was not possible to follow their flight as the rails and sleepers of the bridge were above us.

In the distance we could hear peafowl calling, coming closer and closer. Then their calls ceased, and shortly afterwards we saw them walk into the river-bed. They had come through the bushes. The party consisted of a stately cock and his five wives. All of them drank. Then they walked a few steps down stream when the cock, taking advantage of the open sandy space, decided to display to the utmost his magnificence by spreading out every gorgeous feather in his incomparable tail. Though he stood only on sand, yet he transformed the entire scene into "a peacock throne". While he was displaying his superb loveliness, the members of his harem, overwhelmed by such grandeur, stood motionless in rapturous admiration.

The tail of the peacock is undoubtedly the most beautiful in the world. No wonder that Shahjahan, Emperor of India in the seventeenth century, and a connoisseur of thrones, chose such a tail as a model with which to adorn his own throne, reputedly more splendid than that of any monarch before or after him. His famous "peacock throne" still lives in the writings of historians.

Proceeding a few paces on the sand, the peacock folded his tail, and then with his wives flew towards a large-crowned tree from which they called to all jungle folk, as though to bid them good night. They were the last of the birds to visit the pool. After their final call, a great and awe-inspiring silence fell upon the entire jungle scene.

This sudden silence, and the deep stillness of our surroundings, brought moments of intense emotion impressing upon us as never before the insignificance of man in the overall pattern of the Creation. The coming and going of the birds, in all the splendour of their plumage, was, however, the mildest of introductions to the grim scene that was to follow and hold us spellbound.

A pair of jackals came through the bushes. They must have chosen one of the game paths. For a full minute they stood on the bank, scanning the river-bed up and down, as well as the opposite bank, raising their heads to take in, with their remarkably keen noses, the many scents in the air. They were panting heavily and their tongues were lolling out, witness to the cruel heat and the end of another long summer's day. Stepping cautiously, and still looking in every direction, they reached the edge of the pool and, dog-like, lapped and lapped again. They had to drink copiously, for they had the whole night's hunting before them, and they knew not when they would come for another lap; if all went well, it might be the next morning at the same pool. They trotted along the river-bed under our bridge.

The next visitor was a barking-deer. So sleek, and so graceful in every movement, it approached warily, halting repeatedly, ready to utter a warning bark at the least sign of danger, and to dash off in a succession of bounds. Convinced at last that no danger lurked in the vicinity of the pool, the handsome little creature reached the water; it drank, but at the shortest of intervals raised its head to make quite sure that all was well. Having satisfied its thirst it retreated as lightly as it had come, and we lost sight of it in the cover.

We continued to concentrate on the scrub lining the banks. We had to wait some time before our binoculars revealed a head peering all round with longish erect ears, and a black face. It was a lone hyena. It came on the bank and showed its muddy grey and black stripes. This animal has an extraordinary gait owing to its hind legs being shorter than the front ones. Walking along it cuts a grotesque figure, and gives the impression of having suffered a paralytic stroke in the hinder part of its body. Nevertheless, it is capable of covering many miles every night. It drank, then went its way in search of carrion. There was none of the jackals' nervousness about it.

Throwing their heads forward jerkily, stopping at almost every pace, repeating these movements as they progressed extremely cautiously, a small herd of spotted-deer arrived, comprising a stag with a fine head, and five hinds and two young ones of an age big enough to maintain the speed of their mothers. They were excessively nervous and appeared ready to stampede. The hinds led with the young ones, and the stag was three or four yards behind, holding his head as high as he could. Suddenly they shot away as fast as a flash. It seemed to us that they had not quenched their thirst.

Very soon we knew the reason. A leopard had evidently been stalking them and they saw him just in time. Disappointed, he stepped out of the bush, presenting a magnificent picture, his tawny coat dotted throughout with black rosettes. A leopard is undoubtedly the most beautiful of all the Indian carnivora; he certainly captures the eye for the longest time and claims the most admiration. His coat is truly a most fascinating one, far more so than the tiger's.

This was one of the largest we had seen. He stood on the bank gazing in the direction of the galloping deer, a big meal lost, but a purely temporary disappointment; sooner or later he would make up for it. Having drunk, he evidently intended continuing his stalk, for he quickly crossed over the other bank and, assuming a semi-crouching attitude, went off in the direction the deer had taken, now taking cover, then showing a little of himself when he wanted to look down the river-bed to see how far his quarry had gone. Then he disappeared from sight, but from the last we saw of him there was evidence of his determination to succeed. Of all the denizens of the jungle there is none to match his cunning, his skill and stealth while stalking. He stands in a class by himself.

The leopard and his stratagems were banished from our thoughts as soon as a wild boar strutted out of the scrub. On seeing him, my companion, astounded at the size of the animal, violated our pact of absolute silence throughout the vigil, and whispered: "Oh, what a monster!" I, too, was amazed, for I had never seen the equal of this boar in bulk, height, size of head and tushes (tusks). He was a mighty boar, greyish black in colour, a row of bristles beginning between the shoulders and continuing down to his tail. His thick neck, the pair of sabre-like tushes, and the strength and weight of his enormous body made him the personification of invincibility.

There was no nervousness in his movements and he proceeded straight to the pool where, having satisfied his thirst, he calmly sat down in the water and wallowed and wallowed, dispelling the awful heat of the day, and cleaning his massive frame of irritating parasites. His was a supreme self-confidence born of sheer strength, and the possession of weapons which, once they struck, meant grave injury if not certain death. There he lay in the pool as if it was solely his. He seemed to be proclaiming to all whom it may concern:

> This small pool in sand may one day dry,
> I alone, till then, shall in it lie.

Having cooled himself, he ceased his wallowing, slowly left the water and rolled in the pure sand about four paces away, almost in the middle of the river-bed. He rolled again and again. He was evidently feeling increasingly refreshed. Suddenly he arose, as if an electric shock had gone through him, and stared at the bushes fringing the bank, facing them squarely, absolutely motionless.

After a moment or two, from those very bushes, there appeared the menacing head and shoulders of a thirsty tiger. But to the quenching of his thirst there was now an obstacle. Momentarily daunted, he slunk back amongst the bushes and reappeared about five yards further down. It was easy to follow his movements amongst the bushes by watching the successive positions the boar took up, always facing the tiger head-on. Stripes was manœuvring, realizing instinctively that here was an opponent with whom he could take no liberties. The boar continued to stand firm.

The tiger walked down the bank on to the soft sand, and circled the boar who wheeled to face him. The circling and wheeling continued three or four times until suddenly the tiger charged the boar, striking him with all his might, and, with the greatest agility leaping aside after his blow. The boar met the onslaught by turning dexterously and taking it on the side of his shoulder. In this manner the tiger delivered blow after blow; he was so agile in every limb that he easily avoided contact. His blows told their tale, for the boar was dripping with blood from his shoulders downwards. Whenever the boar showed the least tendency to charge, the tiger would step backwards. There was no doubt that, in agility, the advantage lay with the tiger. Charge, strike and leap away: it was clearly attrition that he depended on for his victory. Nevertheless neither the blows nor the flow of blood appeared to daunt the courageous boar.

This was clearly a fight to the death. We wondered how long the combat would continue. Neither tiger nor boar appeared exhausted. The former, always saved by his agility, was unscathed; the latter an increasingly bloody spectacle, his goriness forcing us to conclude that he would very soon be an exceptional meal for his enemy. While such a thought was passing through our minds, something wholly unexpected and highly dramatic happened. The tiger, in striking and endeavouring to jump away from the boar, either lost his balance and landed awkwardly, or skidded in the soft sand. Seeing his opportunity, and astonishingly quick to grasp it, the boar charged straight into him, and, burying his razor-edged tushes in the tiger's belly, ripped and ripped again, reinforcing his thrusts to the fullest by his stupendous might and enormous weight.

With the greatest difficulty the tiger succeeded in extricating himself and stood apart. In this momentary truce the boar also stood still but ever watchful, awaiting the next onslaught. But . . . it never came. A changed and disfigured tiger now stood before him. He was disembowelled, with much of his entrails dangling low and heavily from his belly. Slinking towards the bank, with his dragging intestines, he ascended it and disappeared in the thorny scrub bordering it.

The bleeding boar stood rock-like; breathing heavily, with heaving flanks, he watched his enemy disappear and continued facing that direction. Satisfying himself that there was no likelihood of further attacks, he rolled repeatedly in the soft sand, on the very battle site. Then rising and shaking himself, he walked downstream, and up the other bank, disappearing into the dark of the forest.

There was no doubt who was the victor and who the vanquished. The tiger's fatal slip in the soft sand followed by the boar's lightning-like, belly-ripping charge was a scene we were never likely to forget. Our ardent hope was that the victor's many wounds would heal, and that he would continue to live as such a worthy and wonderfully courageous fighter richly deserved.

Our unique vigil had ended with all the excitement and

thrills we could have asked for. Our mutual treaty of silence at an end, we made up for it by a full exchange of somewhat breathless comments on the grim fight we had witnessed. Leaving our perch on the abutment we climbed up to the rail track and walked happily in the evening light back to our camp.

Rising early next day, we went to the pool, taking our rifles with us. Stepping up the bank cautiously, we craned our necks from the point where the tiger had disappeared in the scrub, most of which was thorny Zizyphus bushes. "I can see him," said my companion, "there he is, lying on his side." He pointed, and I saw the tiger.

No stones were available, so we picked up clods of earth and threw them at him. There was no movement. Approaching closer, we threw more clods. Still there was no movement. Moving warily, and fully prepared, we went right up to him. He had been dead for several hours.

His progress through the scrub had been brought to an abrupt standstill. Part of his dangling entrails had become inextricably impaled on a recurved thorn of a Zizyphus bush, a thorn no larger than a prickle on a rose bush. Thus impaled, the more he tried to move forward the firmer became the thorn's grip, and the more intense his excruciating agony. He was securely anchored. There he lay down and died.

CHAPTER III

THE MAN-EATERS OF THE GORI VALLEY

As an office "Smugglers' Rock" had many advantages, not the least of which was the magnificent view of the lake below, across whose sun-kissed surface the competing yachts sped like gigantic butterflies. The sight of their graceful sails enlivened the day for me and helped along the grudging hours I was obliged to spend there during the rainy season when touring was out of the question. As Divisional Forest Officer in charge of the Naini Tal district forests I lacked neither interest nor employment, but by the beginning of September I was growing restless; I wanted to stretch my legs again along the forest tracks.

In this mood I should have been less than human not to have welcomed the news of my immediate transfer to replace a colleague seriously ill in charge of the East Almora division. For hundreds of square miles these rich forests clothed the Himalayan slopes, their lowest limit 2000 feet above sea level, their highest the perpetual snow-line. Spreading his maps over the office-table, my chief outlined the job to me with an enthusiasm which at times seemed tinged with envy.

This was just the task I needed. From these forests the Army urgently required all the timber I could get floated down and delivered to Bombay, there to be drawn off as a continuous stream of sleepers to feed the ever-lengthening lines of communication with the Middle East.

The forests to be exploited were scattered as much as seventy miles apart by bridle path, and although I was by now a seasoned campaigner, I could see that this job would call for much skilful organization and some measure of endurance. I thought that was all.

The thirty-six miles from "Smugglers' Rock" to my new headquarters at Almora took my porters three days, and none of us minded the breathing space while I caught up with the arrears of paper work unavoidably left by my unfortunate predecessor. Then followed seven days of ten-mile stages through the most enchanting country, as my camp staff and I, complete with camp office, made our way along the bridle-path to Askote which I had chosen as the centre of operations.

My own organization gave me no breathing space now. By the time the lamps were lit that first evening in Askote I had distributed tree-marking duties amongst five officers. The main exploitation work I planned to get done through the agency of experienced contractors to whom I had given advance notice of my plans and requirements.

The first of these appeared the very next morning in the person of the venerable Rájwár of Áskote. His visit was ostensibly a courtesy call but neither of us expected it to be merely that. This aged gentleman, it appeared, was the father of many sons, all of them honest, industrious and expert in timber handling. I expressed my felicitations and said I should be delighted to meet them. The Rájwár would command them to me this same afternoon. He added that in addition to their natural aptitude for the work they would have at their disposal so much of his humble fortune as might be necessary to finance the valuable contracts I would doubtless be placing in their able hands.

In due course several of these sons appeared together with numerous sub-agents, assistants and supporters. Just which of these were the Rájwár's progeny I never fully ascertained, but after a searching vivâ voce examination I approved of two of them, and in Bikram and Mandhata found excellent contractors. Bikram, a perfect tyrant in the handling and payment of labour, but otherwise energetic and intelligent;

Mandhata by his own activity and the careful treatment of his workers earning their respect, and so making them an invaluable team.

They were but the forerunners of a long procession of applicants, and the battle was by no means done when the names of the successful ones were entered on the books. Far into the night went the arguments regarding rates and conditions. Neither side would have expected it to be otherwise, but it was none the less tiring. Forgetful of the majesty of my surroundings, the dark forests on the cool mountainside, the foaming rivers and the deep night sky, I argued, evaded, side-stepped, advanced; then, having settled on the odd farthing that would be multiplied by many thousands, signed the agreement, paid the earnest money. Necessary work, and not in fact so sordid as it sounds.

Almost, I felt, at the point of my pen the work sped along. The contractors melted, to be replaced by energetic partners who called into being huge quantities of stores and foodstuffs, disposing of them into depots which would ensure the essential ammunition of this sizeable army of workers.

In a very short time the pine forests of Áskote were a vast hive of industry, the sole aim of which was to make a contribution to the winning of the World War. It was gratifying to see so speedily the results of one's labours, and still more so to receive a letter of the highest appreciation from my chief.

Less than a mile to the east of Áskote, the turbulent waters of the Gori river merged with those of the Káli, which formed the boundary between India and the kingdom of Nepál. The Gori had its source in the perpetual snows, and travelled at surprising speed over its steep course with numerous rapids.

My timber operations extended up this remote valley, and it was here that I saw, for the first time, real wild men living in caves. They were very lean and shy, and moved with the speed and silent purpose of a forest animal. In vain did I try to make their acquaintance but they would have none of me. Walking along a bridle path, I suddenly came face to face with one of them, and immediately turned, in an endeavour to accompany him; I prided myself on being a fast walker, but with his spindle-like legs he soon outpaced me, and silently went on his way to his forest fastness. I felt curiously snubbed.

The valley of the Gori was densely wooded. Pine covered most of the slopes, but gave way in places to broad-leaved species under which there was an almost impenetrable undergrowth composed mainly of a shrub with stout, branching stems. This undergrowth was between three and four feet high, ideal cover for all animals.

The day came when, in accordance with my programme of movements, I had to march up the Gori valley to repeat my inspection of the timber operations in progress. The bridle-path passed the confluence of the two rivers. Here I stood for some time, map in hand, reading the surrounding country, admiring the scenery and fascinated by the mingling of the different colours of the waters.

Overlooking the whole scene was the towering and pointed peak of Chiplákot, 13,000 feet above sea level. Round about its summit was the home of the tehr, a mountain goat with an amazing capacity to negotiate the steepest of slopes where a foothold seemed utterly impossible. I made my first acquaintance with it in this locality.

Quite near the confluence there was an expanse of flat ground on which a fair was held annually, the scene of much selling and buying, half of the business being conducted on the basis of money, the other half on barter. Nepál was just across the Káli river, and a few miles further up its course was Dhárchula, an outpost of India, beyond which Tibet came in sight.

The majority of the passers-by were traders. They travelled far with their wares carried on the backs of hundreds of sheep and goats. The daily march, with these tiny beasts of burden, never exceeded six miles; it began usually at 4 a.m. and ended at sunrise. These shrewd men, matchless in business transactions, descended in winter to lower levels in India, where they bought goods with which they returned in summer to the Central Asian Plateau to conduct a system of barter on an extensive scale. Their sartorial appearance, designed for the rigours of the routes they travelled, belied their wealth; they were certainly easier to approach than the wild cave men, and I found that even the prospect of trade made them friendly and communicative.

The pleasant hour, spent inquisitively at the confluence, passed all too soon. The Gori valley called. I went my way along the bridle-path that ran along the left bank of the river and roughly parallel to it.

When there was still a mile left to reach my camp, there was a sound in the undergrowth and my eye caught a movement on my right. Halting, I continued to look up into the forest. After a minute or two I could dimly see what was happening. Coming down with difficulty, staggering through the dense scrub, was a group of men. I could not make out what they were doing. As they came closer they appeared to be in a definite formation. "They are carrying an old log which they will cut up for cooking purposes," I said to the Ranger with me. "No," he replied, "it is not a log, it is something else." "It may be a deer which a hunter has shot," I observed. There was no answer from the Ranger.

But I could see no hunter and no sign of a gun. I continued to wonder what the animal could be. I waited for the men to come nearer. Down they climbed, very slowly, pushing away the stout stems of the bushes with one hand and supporting their burden with the other. When they were a few paces from me I was able to see only too clearly what they were carrying.

It was a dead woman.

As the six men carrying her came abreast of me I saw her long, thick, raven black hair hanging over the leader's shoulder on which her head was resting. The men seemed about to pass me, obviously intending to take her to the village, which was situated between the bridle-path and the left bank of the river, but I felt impelled to ask them to stop. I wanted to know more about this tragedy.

Very reverently they lowered the body and put it down in front of me. She was about twenty-eight years of age and, like all hill women, had a fair skin. Her features were remarkably handsome. From the waist upwards she was naked, the lower part of her body was covered by a short, brown, striped skirt. Her necklace, silver interwoven with beads of various colours, was still round her neck, and the silver bangles on her wrists were intact.

I did not need to ask what had happened. This was the work of a jungle killer. The beast had been able to begin devouring her before being driven away. The left breast had been eaten, and much blood had flowed from it. The other part attacked was the left thigh; a large piece had been taken out of it, and the whole of the leg horribly mangled. In making his attack he had struck and broken her neck.

I asked the men from which place she had been taken. "From there," said one of them, pointing to a spring not more than ten yards ahead of me on the bridle-path itself. An exceedingly bold and hungry tiger, I thought to myself. The man continued: "She was bathing and washing clothes at the spring when she was taken. We were ploughing our fields near the river when we heard her scream. We ran to her help, but we were too late. The tiger had already dragged her away and far up into the undergrowth. We made up a party of six and went in search of her, making a great noise, and following the signs shown by the bushes through which the tiger had taken her. On account of our shouting the tiger left the body."

There were the clothes, washed, and to be washed, still lying near and scattered about the spring. I was very keen on destroying the man-eater at once, and suggested that if the men would put the body back where they had found it, I would go with them and wait for the tiger to return, for he was bound to do so in a short time. I assured them I would not allow him to touch the woman. The man in the centre of the three, on the other side, began sobbing. Placing his hand on the body, he shook his head. In the face of that mute suffering I knew I could not insist, and must let the matter drop for the time being. They carried her away to the village, and I resumed my march to my camp.

During this last mile of my journey I inquired from the Guard of the beat, who was with me, about this man-eating tiger; how long it had been in this particular tract, and how many victims it had claimed. "There were two tigers, both man-eaters," he replied. "One was shot the night before last by the Rájwár's hunter. For many hours he sat in the fork of a tree over a cow which the tigers had killed. At last they came at about four in the morning and he fired and killed one of them." I repeated my inquiry about the number of persons killed. "We do not know," he said, "these tigers are new arrivals but already they have struck up and down the valley, bringing fear to all."

As my camp equipment had preceded me, and had passed before midday the spring where the woman was taken, it followed that the tiger had killed and dragged her away between then and the time of my arrival in the afternoon. Another obvious conclusion was that the tiger, having been deprived of his meal, would be hungrier than ever and therefore bolder.

While these thoughts were occurring to me during the last mile, my eyes were constantly fixed on the dense forest covering the slopes on my right. I was working out in my mind the best method and place for a beat to get this man-eater on the move and kill him. In fact, from now onwards he occupied nearly the whole of my thoughts as, if allowed to live much longer, he would certainly prove a terrible menace throughout the valley. In addition to these humane considerations I had also to think of the timber operations in progress in the same forest. I realized that the man-eater had only to claim one sawyer and the rest of them, numbering scores, would stampede at once, and who could blame them for doing so? The whole of the work would be held up indefinitely. Such was the picture which appeared and re-appeared before me as the hours passed.

As usual, everything in the camp was in perfect order. Each piece of furniture in the tents was in its correct position. Moving camp and re-establishing it—the camping season lasted ten months—had been developed to a fine art to achieve order and the maximum comfort.

My senior orderly dutifully informed me that there was a man-eating tiger in the adjoining forest. The tragedy at the spring occurred after he had passed the spot, but the news had soon spread. I then gave him a detailed account of what I had seen. He said: "I have had the sides of the tent nailed down very securely to the ground. Who can say how far this tiger wanders at night?" I expressed my appreciation of his thoughtfulness.

Shortly after darkness fell, I heard a tiger calling loudly in the forest. He was on the move. I knew he was a male from the shape of the footprints I had seen at the spring, and at the spot where he had stalked the woman. Standing in front of my tent I listened. His calls told me that he was walking in a southerly direction, that the short interval between each call signified he was in search of his mate, and, last but not least, that his headquarters were in the ravine straight up the hill from my camp. He was not more than a quarter of a mile away as he continued calling. Then there were no more calls. His silence proclaimed he was on the prowl, ready to stalk and kill anything moving.

I had dined, and was reading documents relating to the timber operations, when only twenty yards from my tents there were frantic cries of "Sher! Sher!" (tiger! tiger!) All I could do in the pitch blackness of the night was to load my shot-gun as quickly as I knew how, step out of my tent and fire a couple of shots into the air. This, I hoped, would scare away the tiger, and give some confidence to the entire camp, including myself.

In the deep darkness the tiger possessed very great advantages, and, in such circumstances, it was wisest for everybody to remain in their tents. Helpless as I felt, there was nothing more I could do at the moment. To wander about outside the tents would have been tantamount to giving oneself to the man-eater.

After my two shots there was dead silence. I entered my tent and, while doing so, my first thought was for my springer spaniel, Jellicoe. He was my sole companion, my very shadow, never leaving me for a moment out of his sight; wherever I sat, whether at a desk, or table, or in a chair, he would lie near me, taking care that a part of his body always touched my feet.

His bed, three feet square and a foot high, was in a corner of the tent. As I entered he was straining to bring it nearer to mine. I undid his chain and tied it to my bed. Showing every sign of nervousness he jumped up and insisted on lying as close as possible to me. His nose was remarkably keen and had evidently scented the tiger. I never knew him to make a mistake in his all-too-short life of twelve years.

Settling down, I hoped I would have a good night. But continuous sleep was impossible. Cuddled up against me, Jellicoe growled intermittently; this told me that the tiger had not gone away; the morning would tell the whole story. When I did doze off, the pungent and penetrating sulphurous odour from the rushing and noisy waters of the Gori river, only thirty yards away, almost suffocated me. For how long could I continue to cover my face to prevent the overpowering stench from invading my nostrils? But whether from the fumes or self-suffocation, I finally fell into a deep slumber in the early hours of the morning.

Awaking at sunrise my first thought was to examine the

ground round the tents. Sure enough, the man-eater's footprints were on every side; they were traced to the road down which he went in the direction of the spring. Jellicoe, with his nose close to the ground, took in the whole story, lowered his stubby tail and came nervously to heel. The doings of the night told me, beyond all doubt, that the tiger was local and there was every chance of seeing him if a beat could be organized.

I had just breakfasted when the timber contractor's two agents presented themselves before me. They lacked their usual self-confidence. I asked: "What happened last night when one of you cried out 'tiger, tiger'?"

They joined the palms of their hands together, and, looking upwards, thanked their parmeshaw (god) for their escape. One of them then gave his account of the incident: "I was sitting at the door of our hut, waiting for my companion to cook the food. I was looking towards him and talking about our home affairs. He was sitting in the open about three yards from me, preparing our food. By the light of the small bonfire burning in front of the hut, I suddenly saw the tiger crouching and ready to spring on my friend. I cried 'Sher! Sher!' My friend darted into the hut and so saved himself. We closed the door as securely as possible and went to bed without our evening meal."

Both the agents then began to dilate at length about the danger to their workmen in the forest, and how great a loss it would be to their master, the contractor, if the man-eater killed a single workman. It would mean complete stoppage of work, the men would refuse to return, and all the advances of money paid by the contractor would have to be wiped off as bad debts. Finally, they themselves would lose their jobs if work stopped. Very patiently I listened to all they had to say, and assured them I realized that, as loyal employees of the contractor, it was their bounden duty to bring all such matters to my notice.

As our conversation ended the contractor himself arrived

on his sweating pony. He had been busy buying foodstuffs from the neighbouring villages and even from those in Nepál, only a few miles away. Having discussed the progress of the timber operations, the number of railway sleepers sawn up to date, and the number that would be sawn by the date when launching and floating would begin, I told him that I intended beating the forest for the man-eater, and would require the whole of his labour force for the purpose.

At this stage, as though to his cue, the Rájwár's hunter appeared accompanied by the usual motley crew of helpers, beaters and hangers-on. Between them they bore, with evident satisfaction, the skin of a tigress. Spreading it on the ground he said: "The Rájwár sends his compliments and presents this skin to you."

Examining it carefully I satisfied myself that it was the skin of an animal very recently killed. Among other convincing features were the paws which smelt to high heaven; taxidermy in rural areas was distinctly crude. I declined with thanks the Rájwár's kind offer. It would have been highly improper of me to have accepted it, seeing that his sons were serving under me as contractors. Moreover, I had high hopes of soon acquiring one for myself.

I questioned the hunter about his vigil in the tree. He described it as follows: "The tigers had killed a cow. I decided it was a good opportunity to sit in a tree overlooking the carcass, and try and shoot at least one of them. Climbing up the tree before sunset, I settled myself between two branches and waited for them to return to the carcass. Many hours passed. Though I could hear the tigers moving about they did not show themselves. Then, at the first signs of dawn, I saw below me their stealthy approach. Then one came into the clearing followed quickly by the other. Quickly I fired at the leading one, and it dropped. The other turned and disappeared. I continued sitting in the tree waiting for sunrise when I could call to the villagers to come.

Before descending from the tree, where I had spent nine hours, I fired another shot into the tiger lying near the carcass to make sure it was dead. Then the villagers arrived in a large body, and we rejoiced together." Then with a smile betokening extreme satisfaction, he added: "The Rájwár gave me a reward of three rupees." I took the hint.

It was now about half past nine. Everybody had breakfasted, and as many men as possible had been collected. The timber contractor's entire labour force was present, and with the villagers the total number was about eighty. Looking up at the forest I explained my plan and gave the fullest instructions. There was no difficulty in making everybody understand what to do as the forest lay before us like an open book. It covered a slope with a western aspect which began at the highest part of a spur running southwards from the Chiplákot peak. A well-defined watercourse, beginning just below the spur and coming down to the Gori river, divided the tract into two clear sections.

It seemed reasonable to conclude that the man-eater might be in the densest part of the forest which covered both sides of the ravine. This particular part could be seen from the tents and held out the best prospects; I had pinpointed it as the spot where he began calling the evening before. But of course there could be no absolute certainty as to his whereabouts. Our best chance of success lay in this preliminary survey followed by systematic beating of the most likely portions of the forest.

Several of the beaters had axes in their hands, some carried old spears, while the majority shouldered long bamboo staves. They were cheerful and determined, but I noticed among them the dead woman's husband, looking lost and unaggressive. Telling the Ranger that I would go about half-way up the ravine to a position below the dense portion of the forest, I directed him to take all the men to the top of the spur by making a wide detour, and to go silently in order not to disturb the tiger, in case he was where we guessed. On reaching the spur the men were to beat downwards, making as much noise as possible, and the general formation of the beat should be like an inverted V, the object being to keep the tiger within it. The beat was to begin as soon as the spur was reached, since I had a much shorter distance to walk to my position, and would be waiting there.

Strangely enough, the Rájwár's hunter was unarmed. Not content that he should rest on his laurels, I made use of him by giving him my shot-gun with instructions to place himself in the angle of the V, and to fire three shots in the air at suitable intervals to keep the tiger on the move, and so avoid the chance of an attack on any of the beaters.

Accompanied by Uttam Singh, a reliable orderly who carried my second rifle, I selected a well concealed position on the ground below the dense forest. The orderly sat behind me. I faced the watercourse, as the shape of the beat would bring the man-eater down it, or near it, if he was where I imagined him to be.

From his key position the hunter fired a shot in the air and the beat began. The men moved forward, shouting their loudest. After some minutes the hunter fired another well-timed shot, the shouting rising to a higher pitch. My eyes were fixed on the watercourse in front of me and twenty yards distant. Despite my careful planning I felt that I certainly had no more than a fifty-fifty chance of seeing the man-eater. I tried to relax, lest my over-concentration should deceive me.

Then I felt a finger pressed into my back. It was the prearranged signal; the orderly had seen the tiger. It was a moment of joy. Looking slightly to my right, up the watercourse, I, too, saw him.

Apparently undeterred by the hunt, he was walking at his leisure, cocking one ear backwards and then the other, listening to the din behind him. He was not yet broadside on to me.

He paused for a moment and looked back. His attention seemed to be concentrated on the noise behind him. He was slowly approaching a point where the whole of his left side would present the target I desired. Never hurrying, he moved along gracefully over the rocks in the shallow bed of the watercourse. It was good to see his supple form with its powerful forearms, easily capable of breaking the neck of an ox with a single blow.

Looking at him just then I could not avoid the momentary feeling of regret at the idea of destroying such a masterpiece in curve and colour, in muscle and might. But there was no hesitation in the menace of his tread, and my resolution hardened as I watched intently every move of this purposeful killer.

A few paces more, and he showed the whole of his left side, from his nose to the tip of his twitching tail, the swishes becoming more emphatic as his annoyance increased at the vulgar tumult in his wake. The bead of my double-barrel express rifle, with its copper-tipped bullets, was already on him, pinpointing a spot behind his shoulder. It was now a question of seconds, a single second, to avenge the killing of his latest victim, she of the long dark hair whose neck he had broken; a second that meant so much to the future life of the whole valley and, therefore, the most vital of all seconds in that tucked-away spot in the world.

Gently, ever so gently, I squeezed one of the two triggers. The crack of the rifle rang out and echoed round and round the valley. The killer collapsed on to the rocks beneath him. Instantly I rose to my feet. It was imperative that there should be no threat to the lives of the descending beaters, so I emptied the remaining barrel into his prostrate body, the bullet entering his neck. It was not then the question of a trophy, but of destroying a killer of men and bringing peace to the valley. Quickly reloading both barrels I advanced towards him. His head was over one of the rocks. There was no movement. The menace of the valley had been liquidated.

The yelling of the beaters had ceased with the sound of the first shot, and the line halted. I told the orderly to call to the Ranger so that all the beaters could be collected. Helter-skelter down they came, brushing away with their hands the tops of the bushes, leaping joyously over the rock-strewn ground. Breathlessly they all gathered round the tiger. The villagers, being permanent residents in the valley, were touchingly full of gratitude, and each one, in turn, fell at my feet, raised his hands in supplication and prayed to his gods to bless me. I, also, was pleased that the beat had been so quickly successful, primarily for their sakes but also for the smooth progress of the timber operations. Timber for the Middle East where so many lives were being lost in battle, was more than an ordinary responsibility.

The Ranger lost no time in issuing instructions to a carrying party. Stout branches were cut and fashioned into poles. The man-eater was carried down to the camp and his magnificence subjected to the mundane measuring-stick. He was nine feet six inches, and had clearly passed middle age, although there were no signs of physical handicaps. But he and his mate were in a locality where no tiger should have been; they did not fit in, either in territorial configuration or in availability of their normal food; the few members of the deer tribe in the area were difficult to approach, and lived on steep hillsides impossible for any tiger to negotiate. In short, they were out of Nature. To keep themselves alive they had resorted to the easiest of all preys, men and women.

Where did this pair of tigers come from? The natural haunts and breeding grounds of tigers in British territory were over a hundred miles away, in the sub-montane and Terai forests. The only reasonable assumption was that this pair, perhaps driven away by younger tigers, were unwilling immigrants from the Nepál forests across the Káli river, where tigers live in their natural environments and thrive on their normal food. This particular pair had very probably swum the river, and then wandered up the Gori valley. With their destruction I felt reasonably satisfied that our immediate troubles were over.

Completing my inspections of the timber operations, I left the valley by the bridle-path up which I had come. As I reached the spot where the dead woman was brought down through the undergrowth, I saw the bereaved husband waiting for me. He came forward, took off his cap, stooped down, and with both hands gripped my ankles, at the same time pressing his forehead against my feet, and mumbled, with his sobs almost choking him: "You have killed him."

He could say no more. It was one of those moments when a heart is breaking, and words will not come. He looked at me through his tears. I placed my hands on his shoulders as a gesture of my unspoken sympathy, and went slowly on my way.

When I had gone about thirty yards I glanced back. There he was, still standing and looking in my direction, with his hands clasped. A very lonely and sorrowing figure to whom my heart went out, the happiness of whose home had been so cruelly destroyed by the man-eater.

CHAPTER IV

EXPERIENCES WITH BEARS

THE HAMLET stood deserted in the January sun. There was no sign of life, no woman in the house, no man in the fields, not a cow or a bullock to be seen.

As I descended the pine-clad hill on my way to the stream below where I intended taking my lunch, I had a perfect bird's eye view of the little settlement. Here in the Kumaon hills the village houses are constructed in barrack-like formation, the upper rooms occupied by the families and the ground floor by their cattle. The roofs are built of heavy timbers on which are laid massive rough-hewn slates. The effect is one of compactness and security.

But during the four months of winter the families leave their mountain homes and with their cattle migrate to the foot of the hills, known as the Bhabar, to occupy and cultivate their agricultural holdings, and to work in the forests where there is unlimited employment. It is not unusual for a village to be left completely deserted except for an aged caretaker.

Before departing on this annual migration the villagers store their surplus grain in the upper rooms under the heavy slates, safe against their return in the spring. Below me these little granaries stretched on either side of the deserted roadway.

I had covered about half the distance to the stream when my eye was caught by a movement between two of the houses. Into the sunlight emerged two dark shuffling forms. I did not need my glasses to identify these prowlers as a bear with a half-grown cub.

Knowing the habits of bears, and suspecting why the

mother and her cub were visiting a deserted village, I halted some little distance away, sat down, and with the assistance of my binoculars became an interested spectator. With me were the Ranger, several Forest Guards, and an orderly, whom I bade sit down a yard or so behind. My instructions to them were to stay perfectly silent and to be careful not to dislodge loose stones which might roll down to the village and disturb the intruders.

The bear, with the cub hard on its heels, walked between the houses, now and then stopping and sniffing the air. Her sense of smell seemed to reveal a secret. Advancing to a conveniently situated tree she climbed it and the cub followed her. From the tree she jumped on to the roof of a house. The cub was no less agile. An interesting scene developed. She walked along the roof with her nose well down. Eventually she came to a stop over a particular room and promptly began lifting and rolling over a few of the slates, each of which I knew to be about two and a half feet long, a foot and a half wide, and an inch thick. She and her cub were obviously desperately hungry. It was mid-winter and fruit on trees and shrubs was scarce.

She succeeded in moving four or five slates. Having done this, she and the cub disappeared into the room. From the men sitting near me came cries of "Thieves! Thieves! They will eat the grain." I told them not to say a word, that all would be well. The orderly was seated closest to me with my rifle, but I had already made up my mind not to shoot the hungry bears. I was greatly interested in them and preferred to await developments.

The sun was not yet directly overhead. There was plenty of time in which to have my lunch and ascend the opposite hill where I had to inspect the oak forests. I therefore continued to focus the binoculars on the hole in the roof made by the starving mother bear. Since in these matters her nose would be infallible I concluded that she and her cub were filling their stomachs as quickly as they could, knowing that the time factor was all-important lest they should be disturbed at any moment and driven away unsatisfied.

On to this scene, as if to a cue, appeared the guardian of the village, a much-bent old woman making her way to a clean section of the stream to fill her can. As she reached the burgled house she stopped, and began screaming at the top of her voice: "Robbers, robbers!"

Then as fast as her spindle-like legs could take her she climbed to a point about twenty yards on the opposite hill and began to shriek to the caretaker of the neighbouring village for help. She then picked up her can and valiantly returned to the house. Taking a long stick she beat the walls to the accompaniment of all the abusive words in her vocabulary. Evidently concluding that her outburst had not the slightest effect, she came to the front of the house, climbed the steps and reached the front door. She tested the lock and found it intact. Then she put her ear to the door, and kept it there for two or three minutes. "There are bears inside," she cried, "there are bears inside, they are eating grain, they are eating grain." With this the poor old soul became hysterical, wept as loudly as she could, and collapsed several yards from the house.

The time had now come for amusement to give way to helpful action. We called out to the woman not to be afraid, and that we would soon be with her. She looked up at us, rose from her prostrate position and waved excitedly, at the same time shouting: "Come soon! Oh, come soon! the bears are inside Guman Singh's house eating all his grain. What will he say to me when he returns?"

We hastened to her and made straight for the house. With my gun loaded I crept up the stairs to the front door. I fired a shot in the air and listened. There was much grunting and hustling in the room, and then through the hole in the roof the bear and her cub emerged, scampered along to the tree, hastily descended it with a jump to the ground and skedaddled up the hillside on which we had been sitting. The mother bear instinctively adjusted her pace to suit her cub's. Away they went up the hill and out of sight. Although they passed a few yards from me it would have been cruel to shoot them. They were obviously in desperate need of a meal, and the pangs of hunger had driven them to adopt the role of burglars.

The aged woman, with "a furrowed brow wrinkled with care", needed much comfort. I told two of my men to go through the hole in the roof and assess the damage. They were let down gently. These roofs are very low so that one is obliged to walk in a stooping attitude. Some minutes later, the men appeared to state their estimate of the cost of grain eaten. In order to be on the safe side I doubled the amount, and added a generous sum for the solitary caretaker. I asked her to hand over the cost of the grain to Guman Singh and to give him my compliments when he returned in the spring. Before saying goodbye to her, we laboriously restored the slates to their correct positions.

The two kinds of bears I have known fairly intimately are the Himalayan black bear and the sloth-bear.

The black bear is so called to distinguish it from the red species which is usually found in a higher zone of the Himalayas. When viewed from the front there can be seen two distinct white marks; the more conspicuous is the one on the chest, resembling a large horse-shoe upside down; the other is a daub on the chin. This bear's claws are black, relatively short and very strong. The coat is otherwise jet black, thick, with hair nearly three inches long; the ruffs have longer hair. I have met some black bears of immense weight with broad heads, sleek coats and bodies remarkably well groomed. They were exceptionally powerful, and altogether fine specimens of their kind.

The food of the black bear consists chiefly of the fruit of forest trees, such as the acorns of the low-level and middle-zone oaks. With avidity it eats the fruit of the many species of brambles growing in the hills. In deodar and blue pine forests it is sometimes responsible for extensive damage; with its teeth and claws it removes the bark of young trees to lick the resinous surface of the naked wood; I have often observed the entire bark removed round the stems resulting in the saplings dying in groups wherever the bears' activity was concentrated. These bears are constant invaders of fruit gardens and are, therefore, anathema to orchardists.

Walking through the forest one afternoon I was greatly intrigued by the shambling figure of a black bear—fortunately at a reasonable distance in front of me—which suddenly stopped, raised its head, and worked its muzzle inquisitively around. Then, without the least hesitation he swarmed up through the maze of branches above him until he was level with what I could now discern was a large beehive suspended from a branch.

I took up a discreet observation-post, and watched him as he dexterously balanced on his hind feet and, reaching out to the hive, attacked it ravenously. Pandemonium broke out. The enraged bees swarmed over every part of his head and coat, and clearly made their impression on his nose and lips, for between grabs at the hive he slapped vigorously at them, and then scooped up a mouthful of honey, wax, and bees. Undeterred by the avengers he made the most of his time, and then descended, stopping now and then to shake his head and wipe away from his face the determined, clinging bees. Reaching the ground he quickly made for a grassy patch to try and conceal himself from further attacks. With the exception of a few small negligible pieces of wax which fell to the ground he had gorged himself on the whole of the large hive. I tactfully left in the opposite direction.

I do not think the Himalayan black bear can fairly be described as an aggressive animal. Like most wild animals, it will justifiably defend itself when cornered. On most occasions I have known this bear to do its best to escape as quickly as it can, even jumping from a high branch to do so, or clambering down the stem of a tree at top speed to get away to safety while I was standing only a few yards away.

A she-bear will very naturally defend her cubs. I saw an unique case when I was inspecting resin-tapping operations in the Thua block in the Naini Tal district. A bear, with her two small cubs, was about twenty-five yards ahead of me on a steep hillside. As soon as she became aware of my presence, she hurried her family along and disappeared into a cave. I was curious to see the cave, so passed close to its mouth which was just big enough for the bear to enter. To hide herself and her cubs, she blocked the mouth of the cave as much as she could with her hind feet. All I could see were the two soles, side by side. The temptation to get a twig and tickle them was almost irresistible. As I was unarmed, discretion dominated mischief, so I continued inspecting the resin-yielding channels on the pine trees, safely away from the dark occupied cave.

Now and then, however, I have come across hillmen who have been mauled by Himalayan black bears, and who showed several nasty scars on their faces, arms and hands. There are two circumstances I know when this species of bear will charge, or, in its sudden mental confusion rush forward and attack. It will occasionally charge when wounded, and if the hunter concerned happens to be below it on a hillside. I have experienced such an event, and I can testify that the charge is fast when it does occur. The other circumstance is when a bear on a narrow mountain path unexpectedly meets round a spur a man walking in the opposite direction.

Such an encounter was experienced by one of my Guards while patrolling the forests in his beat. He was tall, with a straight figure and in the strength of his late thirties. The incident occurred on a lonely bridle-path, known as the Sherring Road, which followed a conspicuous ridge, leaving

High level forests. The home of the Himalayan black bear

If captured young the Himalayan black bear can be trained to wrestle; (left) the challenge; *(below, left)* locked in combat; *(below, right) victory*

the main district road at Mornaula, 7,000 feet above sea level. Winding its way in a general easterly direction, it gradually descended towards the famous old shrine of Punagiri.

In winter hardly a single human being was to be seen on it. Bears frequented it, so did man-eating tigers which found it a convenient route to the lower parts of the Ladhia valley where they carried out their depredations amongst the villages in that locality. It was never safe to traverse the Sherring Road without a gun. The dense oak forests on both sides caused visibility to be poor, and added to this disadvantage, the road went round numerous spurs, thus blocking the view ahead.

In such circumstances the Guard suddenly met the bear, and the two of them almost collided. Instantly standing up, it began to strike at his face, which he instinctively tried to defend by raising his arms. Recovering from the muzziness of the initial shock, and able to think clearly, he remembered the modest little weapon suspended from his belt. This was a combination of axe and hammer with a handle a foot long; on the hammer was inscribed his official serial number. Continuing to keep his left arm raised to ward off the blows, he succeeded in releasing the weapon with his right hand. He then brought it down, as hard as he could, on the bear's nose, an extremely sensitive part. This counter-attack had an immediate and salutary effect. Subdued, the bear dropped quickly on all fours and shambled downhill into the undergrowth, there to feel the after-effects of the encounter.

But the Guard also had his mementoes of the encounter. Fortunately his coat and shirt, although torn in shreds, had saved his forearms from the worst of the mauling strokes. Although suffering from shock he pulled himself together and accomplished the long six-mile trek to my camp. Washing his many deep lacerations thoroughly, I applied permanganate of potash, bandaged his hands and gave him a dose of brandy. With two other Guards to keep him company, I sent him to headquarters where he was admitted immediately as an in-patient. He was discharged after about three weeks. The palms and backs of his hands had long and ugly scars, though curiously little injury was done to his fingers, and, also luckily, his face was unscathed. He owed his life to the presence of mind which prompted a telling blow on the right spot. History does not relate whether his adversary bore the tell-tale number on his snout.

Provided the wounds are treated within a reasonable time, and have not been inflicted to a fatal extent as sometimes happens when a human victim collapses on the ground and is at the complete mercy of the bear, they are less likely to turn septic than are those caused by a leopard or tiger. These animals invariably carry putrefying flesh in the inner surfaces of their claws, which, unlike those of the bear, are retractile and harbour such rotting matter permanently; it is this matter that often gives rise to serious blood-poisoning when a man is attacked.

Against this, however, the helpless victim of a carnivorous bear suffers most cruelly, since the end is not quick as is the case when a leopard or tiger kills. The bear mauls, and continues to maul, at the same time taking chunks of flesh out of the struggling prey. This is particularly true in the case of death caused by a sloth-bear, which is in the category of the worst of all deaths. It is characterized by a terrible disfiguring of the face and body as the merciless claws dig deeper. I have every reason to regard such a fate with the utmost horror.

I was one of several officers selected to undergo a special post-graduate course in Advanced Forestry, and a vital part of the syllabus was tree-selection under the several kinds of Silvicultural Systems. It was a procedure for testing an officer's knowledge in the branch of Practical Silviculture, and so demanded judicious decisions in marking trees for felling, consistent with the well-being of the forest as a whole.

The varying sets of conditions met were exceedingly complex. The test, once passed, was therefore an enviable professional qualification.

We made extensive tours throughout the United Provinces to forests of all types and amongst these were the sál-covered areas of the Eastern Circle, known for their excellent quality and great value. Arriving at our destination by train, after a tedious journey, we settled down in our tents. The following day our professor assigned a plot to each of us, and we began tree-marking immediately. I was allotted plot number five with instructions to mark six hundred trees, each of which was to be over two feet in diameter. The selection of the trees was governed by prescribed conditions. A time limit of ten days was fixed for this work.

Accompanied by a Guard and a gang of workers I reached my plot daily at nine o'clock. These workers were employed in different ways. Two were termed linesmen; separated by about twenty yards they went along in successive trips within which I selected trees. Another two were known as axe-men. Their role was to blaze the trees at ground level and at chest height. They were followed by a man with a special marking-hammer bearing certain letters. Next followed a worker who put on the serial number of the marked tree. These operations having been completed, a worker applied a calliper to the tree at chest height and called out its diameter. This latter measurement, as well as the serial number of the trees, was recorded by me in the tree-marking register. At intervals the sequence of numbers was checked. The Guard's function was to keep a vigilant eye on all the workers, and to see that each played his part efficiently.

I moved slowly from linesman to linesman, studying each tree and giving my decision. In the register all entries were made in pencil and inked in daily on return to the camp. Eighty per cent of the trees were sál, the remainder comprising those species which ranked below in value.

Wherever the forest canopy was broken, intense light

entered and resulted in patches of dense grass, three to four feet high, from which a deer would occasionally rise and dart away. Now and then there would be a silent, stealthy movement as a tiger walked to another plot of grass to continue his siesta. There was reputed to be a complete absence of man-eaters in these very attractive forests.

As the operation continued day after day the possibility of being attacked by animals receded completely from my mind. Moreover, the local conditions were entirely new to me and bears were never further from my thoughts. My sole endeavour was to pass the test and to complete the work within the specified time.

On the sixth day, I was compelled to continue work beyond five o'clock in order to complete a strip in hand. By this time the light in the forest had diminished. Finishing the strip, I collected the workers, and was leading them through the forest in the direction of the camp, walking at a brisk pace and competing with the fading light to avoid being benighted.

Occasionally, we passed through grassy glades. As I entered one of these my eye caught a movement on the edge at about twenty yards from me. Turning fully towards it, I was horrified to see three bears standing up on their hind legs and looking at me. The one in the centre stood about five and a half feet in height, while the other two were somewhat shorter. I formed the quick impression that they were a mother and two almost full-grown cubs.

Halting, I pointed them out to the Guard.

"Three bears," I said, with simple significance.

"Yes," he answered.

No sooner had this revealing conversation taken place than all three came charging straight at us, grunting a chorus of murderous intentions. They presented an excellent target; but alas! my only weapon was a lead pencil! I did the quickest thinking of my life. To run forward I should find the grass a treacherous impediment; the determined enemy would overtake and fell me to the ground to give full vent to their fury. A lightning decision made me turn and run in the direction from which I had come.

"Run and save your life," was the only advice I could give the Guard. That was the last I saw of him.

Turning quickly, I began to run. I had not gone five yards when, in my eagerness to escape, I stumbled over a large log hidden in the grass, and fell flat on my chest. I felt all the air go out of my lungs. My hat dropped off, and the right lower pocket of my coat was ripped away.

The fall left me gasping for breath. As I lay trying to recover, I felt the ground vibrate with the weight of my pursuers. "This is the end" were the inevitable words that flashed through my mind. I expected all three bears on me, scratching and digging into me as I lay prostrate and breathless, completely at their mercy.

Seconds went by, my terror grew. And each second seemed an hour. Then I realized I was still safe. Due to sheer good fortune the bears had passed by within a few feet of where I lay, partly concealed by the grass and partly by the once treacherous but now friendly log.

Unscathed, I rose with difficulty and could see neither man nor animal. The feeling of still being alive was altogether too wonderful for words. Hatless, I ran back and saw the men standing at a safe distance. Not seeing the Guard, I asked where he was. "There he is," they replied, pointing to the branches of a tree. We kept our eyes on him, he occasionally glanced at us when not observing the ground below. He made signs which told us the bears were still there.

After some minutes had passed, he called out: "They have gone", and, coming down from his perch, joined us. He explained that when the bears came on at speed, he made the mistake of running forwards in the hope of reaching the other side of the glade and so outdistancing them. But the bears' pace was too much. He then decided to climb the first tree he could reach. It was during these brief moments of indecision that he was in full view of the bears which consequently decided on making him the object of their attack, since I had disappeared from the scene by falling over the log into the grass. While climbing the tree, his turban, which was part of his uniform, fell off; the bears in their fury tore it to shreds.

We waited; then returned very warily to the place where I had fallen, and recovered my hat. Very fortunately, the precious tree-marking register was safely folded in my inner pocket. I decided to abandon the lead pencil!

We resumed our journey through the forest, but decided to change direction in order to emerge on the nearest fire-break; how far it was the Guard was reluctant to guess. However, we walked as we had never walked before. In addition to being spurred on by the thought of meeting more bears, we found that twilight had set in, for it was the depth of winter, and the days were the shortest in the year.

A sight of the camp was definitely consoling. On reaching my tent, I dismissed the men and was glad to sit in a chair with peace all round me, for I was not a little shaken. My neighbour, Crothers, dropped in as usual to compare notes relating to the day's work. I described, as best I could, the events of the last hour. "A very narrow escape," said he, "and this must be your lucky day! We have just been informed that it is dangerous to be unarmed in these forests. We have been advised to take our guns tomorrow and every day thereafter."

This was superfluous to me; I had already made up my mind to do so.

Next morning, when our professor heard about my experience, he decided to accompany me throughout the day's work. He brought his magazine rifle and circled round the plot, while I continued to select and mark trees with my gang of workers. He kept within earshot but did not see the bears. Early the following day he made a special excursion in the hope of meeting them, but fortune was on their side and not on his. I completed my task with a day to spare. I decided I could not spend it in a better way than to try and meet my enemies. I searched every patch of grass. Alas, my day-long hunt was in vain.

But the Fates were kind to me. Eleven years later the Government posted me to the charge of the very forest division in which I had had such a narrow escape. Accordingly, I took the fullest revenge on sloth-bears, and killed over sixty in those forests. I looked upon them as a dangerous pest. They had multiplied greatly, and it was imperative to reduce their numbers.

The sloth-bear, also known as the Indian Bear, lives in dense as well as sparse forests on the plains and low rocky hills. Its habitat begins at the foot of the Himalayas and extends over the whole of the Indian peninsular. Black, with a coarse coat made up of long hair it never looks clean or sleek in appearance. The claws are dull white, the muzzle a muddy grey as if it were in urgent need of a thorough scrubbing with soap and water. Very highly developed are the snout and lips.

The average size of the numerous sloth bears I have killed is not equal to that of the Himalayan black bear. Of course there are exceptions, big specimens of both kinds being met from time to time in favourable localities. Usually there is more fat on the mountain species, a protection given by nature to withstand cold and snow. This fat is prized by Indians as a cure for rheumatic complaints but, in efficacy, ranks second to a tiger's.

The mention of rheumatic complaints brings back to my mind the scramble that used to occur among the máháwats as soon as I killed a male sloth-bear. They jumped down from the necks of their elephants, or climbed down their trunks as quickly as they could. The one who reached the bear first, turned it over, with the belly facing upwards, and proceeded with a hunting knife to cut out its sexual organs.

I inquired the object of this operation but the only reply I received was "for medicinal purposes".

Sloth-bears, which spend their lives in forests with heavy undergrowth and patches of grass, suffer severely from pests, the worst by far being a species of tick. On many bears I have shot I have found these ticks, bloated with blood, hanging like bunches of grapes on a vine, from the neck, chest and armpits, a truly revolting sight. In the rainy season, when ticks are less numerous, the bears are bled by leeches which, when satisfied, are as thick as a man's thumb and half as long again. Thus these animals carry blood-demanding parasites at all seasons of the year.

Its food consists mainly of fruit, roots and insects. As a climber of trees it cannot match the skill of the mountain bear. I have observed it moving most awkwardly amongst the branches of trees while endeavouring to reach a bee-hive for honey. In sál forests this bear's speciality appears to be white-ants which are found in their conical-shaped nests laboriously built up to a height of four to five feet above the ground. These ant-hills are scattered throughout the forests, even in open situations like a hundred-foot-side fire-line.

To satisfy myself, I carried out a cursory survey of such ant-hills in order to see what percentage of them had been visited by bears. I did not find a single one intact. They had all been burrowed into at ground level. In each of them there was a hole about nine or more inches in diameter through which the bears had pushed their muzzles and sucked up as many termites as possible. Their powers of smell and suction are truly extraordinary.

Sloth-bears also enjoy eating the petals of a certain flower which blooms profusely in March and April. Late in my career, I was appointed to the charge of Jhansi forest division, the southernmost administrative unit of the United Provinces. The boundaries touched Central India, several petty States and, for a part of their length, coincided with those of the Central Provinces (now called Madhya Pradesh). The forests, though interesting in their own way, were very different in type from those in the Himalayas and submontane tract. The trees were often widely spaced, the stems short and the crowns large and spreading producing prolific crops of flowers in spring.

One such species was the mohua (*Bassia latifolia*). With the ripening of the ovary into fruit, the petals are shed and fall to the ground, covering it like a carpet. They are cream-coloured, almost an inch long, very pleasantly perfumed, fleshy and sweet. The local inhabitants eat these petals either raw or cooked, regularly taking them to the bazaars for sale to form an important ingredient in certain sweetmeats. In times of famine these petals are dried, and ground into flour from which a type of bread is made; they are also distilled to yield a coarse and highly potent spirit.

On one occasion, effectively concealing myself, I watched three bears approach these mohua trees and enjoy an exceptional feast. They arrived together in their usual social manner, and lost no time in greedily beginning on the petals. Their appetites seemed insatiable.

Then came a most laughable sequence. All three were unmistakably intoxicated. Slowly departing, the revellers moved drunkenly, stopping now and then to balance their swaying bodies. One of the trio, smaller than the other two, was so much "under the influence" that it found temporary comfort in a prostrate posture, rising with difficulty when it dreamily realized the fear of being left alone. In these circumstances, despite my very strong feelings against sloth-bears, I felt it would be unsporting and heartless of me to shoot the happy inebriates.

The sloth-bear can be a noisy animal. Two of them disturbed my slumber for the greater part of a night when I was camping at Changa-Nala bungalow in the North Kheri division. On three sides there was dense forest only thirty yards away. I retired at ten o'clock, but an hour later was awakened by a succession of loud grunts on the edge of the compound. Listening, locating and recognizing, I lay awake hoping the noise would cease. Vain was the hope. The din continued and showed no signs of abating. I tried to forget it, but it proved too loud. Sleep was out of the question, and by my watch it was well past midnight. Something had to be done to bring about silence and slumber in the interests of the day's work ahead.

Changing my night-suit for a pair of khaki shorts and a khaki shirt, and putting on a pair of rubber-soled boots, I loaded my double-barrel rifle. With a powerful torch in my other hand I walked to the outhouses to rouse an orderly. But there was no need to wake him. He, too, complained about the noise in the forest.

I explained my plan, and instructed him to be careful to stand close behind me, to enable him to throw a beam of light over my right shoulder along the barrels of the rifle at the bears. We then crept to the edge of the forest towards the noise. The bears were busy concentrating on each other. Their loud and fierce grunts were in our favour. Creeping towards them, and making sure the orderly was in the right position, I raised my rifle to my shoulder. My companion responded correctly, and my aim was soon on the contestants. They were standing and striking at each other as fiercely as they could. In their erect position they made good targets. Aiming first at the bear on the right, and then at the other, I killed both of them. Leaving them lying where they fell I went back to bed. I examined them in the morning. They were two good-sized males, and had joined battle either for the possession of a she-bear or to settle a long-standing feud. Whatever was the cause, the result was a settlement in which neither party won.

CHAPTER V

ELEPHANTS

KING CHANDRAGUPTÁ THE FIRST, who ruled over nearly the whole of India three hundred and twenty-two years before the birth of Christ, employed no less than 6,000 elephants in battle. On the day of the combat they were arrayed in coats of mail armour, their tusks being equipped with spikes to add to their effectiveness. Such a colossal force, ready to advance on the enemy, defies both description and imagination.

In India of the present age, the unsurpassed skill and strength of elephants are used in timber operations, and more spectacularly they are employed during marriages, festivals and other important occasions, when processions pace slowly along the principal roads of a city, led by these majestic animals, gorgeously caparisoned with the rich and colourful trapping. For sheer splendour and dazzling brilliance these processions are surely unrivalled.

A common purpose for which these amenable animals continue to be used, is for hunting tigers, bears, and leopards. Apart from sport, however, elephants are indispensable in facilitating inspections in forests when swamps and wide shallow rivers make it impossible to negotiate a tract on foot.

Prior to 1947, when the political and national status of India changed with dramatic suddenness, every large estate owner, rájáh and máhárájáh, maintained elephants, the number depending on his annual income, as well as on the degree of dignity demanded by age-old prestige. The periodical replenishment of these numbers was carried out by forced recruitment from wild elephant herds.

Such an operation is known in upper India by the word *kheddah*, which appears to be derived from the Hindustáni verb *khedna*, in common use and meaning "to chase". It is an organized pursuit to capture wild elephants; the mode of capture may be by driving them into large pits, or stockades, or chasing and then surrounding and roping them, the pursuers being on tame elephants.

Where the methods of pits and stockades are not adopted, the capture of wild elephants, especially those with tusks, involves a prolonged combat, preferably on fairly open ground. The procedure is to locate the tusker and then bring him to battle, pitting against him, in succession, several tame tuskers, till the wild one is eventually exhausted and defeated. He is then shackled with heavy chains, tightly sandwiched between two tame tuskers and thus made to walk to the camp, there to be securely tethered by all four legs, the iron chains used being particularly strong since strenuous efforts are made to regain liberty. They pull and pull so much in their endeavours to get free that I have seen the skin on their hind legs, below the knees, hanging in shreds.

The first large-scale elephant-catching operation of this sort I witnessed was by about forty elephants belonging to the Máháráját of Balrámpur, an estate with an enormous revenue, and the biggest in area in the United Provinces. The scene was the southern slopes of the Siwálik range of hills, in the Dholkund forests of the Saháranpur district. The Máháráján, with his elephants, camped at Beribára, only six miles away to the east.

The Máháráján's men had isolated two magnificent wild tuskers. They were enormous, of exceptional height and with long pointed tusks, altogether splendid specimens. It took all the tame tuskers to defeat the first one, and by the time the second was engaged the attackers were tiring. Time after time one of them would be withdrawn, to be quickly succeeded by a fresh one ridden into the battle by his bold máháwat. But the wild tusker was more than holding his own against his several adversaries and appeared to be invincible. It soon became evident that there was considerable doubt regarding the outcome of the struggle.

The Máhárájáh shared this anxiety and called repeatedly for his champion tusker: "Where is Bob Báhádur? Where is Bob Báhádur?" For some reason best known to his máháwat, Bob Báhádur had, on this occasion, lagged behind. Horsemen were sent galloping back in the direction of the camp with stern orders to his máháwat. That stalwart, in fact, needed little prompting. He had no doubt purposely withheld his charge until the critical moment. Bob Báhádur was on his way and being urged forward as fast as his legs could carry his huge body.

On he came, holding his head high, while on his neck his máháwat shuffled from side to side digging his puny heels and toes into the impregnable hide to convey his message of urgency. Bob Báhádur looked every bit a champion. He seemed to know, by the poise of his head, why he was being urged. Arriving on the edge of the arena, and seeing the combat in progress, he went straight in and joined the fight against the wild tusker who had just routed his latest opponent.

Then began a battle between two giants, veritable mountains of flesh and strength. Clash followed clash, thrust after thrust, the tusks meeting and clattering loudly. The Máhárájáh's expression of anxiety gave way to smiles of optimism as he looked on and saw distinct prospects of an exceptional capture. Bob Báhádur was battling away determinedly at the wild tusker. It was not the first time that the Máhárájáh's favourite elephant had fought for him. We marvelled at the wild tusker's endurance, but it was clear that it could not last indefinitely against such odds. Finally he fell, completely exhausted but fighting to the end. The nearby máháwats raced to be the first to shackle him and thus gain a prize at the hands of the Máhárájáh. This was no quick task, but eventually he was prodded to his feet, and, closely sandwiched between two of the tame elephants, taken triumphantly to the camp. Bob Báhádur, looking rather like Mount Etna after eruption, shook his mighty head and stalked off on his castle-like legs; last in, first out, the old campaigner lived through many more such battles.

The training of the newly-captured tusker was an interesting spectacle. The máháwats engaged in this education were fearless men. In the company of two powerful male elephants he was taken out daily. Literally, he was wedged in tightly between them. Whatever they did, as commanded by their drivers, he was, by their combined weights, compelled to do. The necks of all three animals were so roped together that, from a front view, they had the fearful and fantastic appearance of a legendary monster, a kind of heraldic horror. Whenever the captive showed the slightest stubbornness, there was another tusker behind to administer, with a fierce prod in his hindquarters, a persuasion that could not be ignored.

In this anything-but-gentle manner the trio, with the necessary reinforcements in the rear, progressed daily. Very naturally, the wild one would repeatedly display his strong resentment at such treatment and bondage, but he was held firmly and there was no possibility of escape. A forward bid-for-freedom dash was out of the question since it was beyond his strength to drag two monsters, each equal to him in weight. When he attempted such a movement the máháwats gave their mounts the order to resist, which they did either by standing still, or backing. It was a continuous performance of counter strains and counter stresses to inculcate obedience, through words of command in a language which, the sooner he picked it up, the better for him. On all sides during these lessons he was surrounded by brute force and intelligence.

In due course a máháwat sat astride the wild one's neck to teach him the meaning of certain signs made by toes and heels behind his ears, and well known to all domesticated elephants. All three máháwats would simultaneously make the same sign, and shout the same command. If the word was "baith" all three would sit down. At the sound of "mile" all three would stand up. This training continued for about six months, after which the captive qualified to take his place amongst the Máhárájáh's herd as an obedient and respectable member.

Taking thirty elephants of his herd the Máhárájáh made a wide sweep, a few days later, through the Dholkund forests. News had been received that many wild elephants, mostly female, had taken refuge there. Arriving very early in the morning the operations began. By three o'clock in the afternoon no less than thirteen had been captured; amongst these some had babies not quite four feet high.

The progress of captors, with their captives, on their return journey to camp, six miles away, was exceptionally interesting to watch. The expert máháwats went about their business as if they were rounding up a herd of farm-bred cows. I observed a mother with her baby being shepherded along. She was in the middle of four tame elephants, each tied to her by ropes ten feet long. The four were spread out; one to the left front, another to the right front, a third to the left rear and the fourth to her right rear.

The pace was that of the baby. Maternal anxiety was intense as the little animal showed a tendency to stray. Out would stretch the mother's trunk gently to draw the would-be wanderer to her side and occasionally under her belly.

If the mother attempted to rush forward the two rear elephants would pull backwards. If she made an effort to retreat the two front ones would tug their ropes forwards. When she refused to budge there was a spare elephant which was immediately driven forward to butt her rump and make her go on.

The organization and highly efficient day-to-day working of the Máhárájáh's camp was nothing short of miraculous.

He had undertaken these elephant-catching operations at the special request of the Government, who desired that the number of wild elephants in the province should be reduced, since they had multiplied so greatly that they had become a menace to villagers' crops in three or four districts. Up to the time I bade the Máhárájáh goodbye the total number of captures was thirty, including babies, but there was still a very fine specimen of a tusker the Máhárájáh was keen to capture; it had had the audacity to visit the camp the night before and actually entice away one of the newly-caught babies. We wondered why a huge tusker wanted a baby to keep him company, and could only conclude that he was hopeful the mother would free herself and join him and her offspring.

In a forest, as elsewhere, incidents verging on the tragic occasionally end on a humorous note for the onlookers. At a time when malaria was rampant amongst my staff and labour, I obtained official sanction to establish a travelling dispensary and to appoint an Assistant Surgeon in charge of it.

When this newly-arrived man of medicine saw nine elephants lined up in the camp for a tiger hunt, he advanced briskly towards me as I emerged from the rest-house and said excitedly: "I have never seen a lion. If you are going hunting today please allow me to come as I am very anxious to see a lion."

He was entirely town-bred and appeared to think that a lion and a tiger were one and the same animal. "Certainly," I replied, "get up on one of the pad elephants." Then, turning to one of the máháwats, I bade him take the doctor on his elephant.

We went to a part of the forest where there was an island of tall grass, about three miles from the camp. I placed everybody in position, with the howdahs on the flanks, where the tigers were most likely to appear, and then, as

A line-up in the plains forests for the day's hunting with emphasis on sloth bears

(Right) A family of sloth - bears shot during the expedition

The author photographing a wild male elephant which, in its rage, covered him with spittle

usual, the pad elephants were directed to go through the grass and disturb the animals.

Moving slowly along we came upon a tigress with two quarter-grown cubs. This was hunting, and quite correctly it was against the law to shoot such a tigress. She and her cubs trotted along in front of us. Not liking the situation, she ran back through the line of pad elephants. Realizing that her cubs were no longer with her, enraged at missing them, and holding the elephants wholly responsible for the cruel separation, she made up her mind to turn and give vent to her fury. The elephant she singled out was, unfortunately, the one on which the Assistant Surgeon was seated.

I saw her leap right up, with her forepaws dug deeply into the pad itself, snarling open-mouthed as only a tigress can. The doctor quickly changed his position and retreated as close as possible to the máháwat, who was doing all he could to control his panic-stricken elephant which was feeling the claws of the tigress's hind legs more and more in her hind-quarters.

Not unnaturally the Assistant Surgeon appeared petrified. The tigress's head was not more than three feet from his own. He was holding on to the ropes of the pad for dear life, and the elephant was aggravating the situation by continuing to panic.

All that separated the doctor from the tigress was a pair of old spotted-deer horns the máháwat had told his elephant to pick up from the grass, and which he had tied on the rear portion of the pad just before the incident occurred. The horns were making it difficult for the tigress to advance further on to the pad and seize the doctor, but, seeing his very real peril, I quickly put my rifle down and, picking up my shot-gun, fired into the air over the tigress. She immediately slithered down and disappeared.

I hastened on my elephant to the trembling medico. He was beyond uttering a sound, let alone being able to speak. The colour of his face was a peculiar mixture of ashiness and a ghost-like pallor. I had every sympathy for him as I helped him on to my elephant to assist his recovery from a very frightening ordeal. When he had sufficiently recovered to speak, he said, in the lowest of tones: "I will never come again; I could even smell the lion's disgusting breath."

Buying a horse at a fair is notoriously a mug's game, which I have so far avoided, but I once went to a famous fair and bought an elephant. I selected her, sound in wind and limb, strong of tusk, and with an eye as gentle as a Jersey heifer. I paid for her, rupees four thousand five hundred, on behalf of the Government, and thought they had a bargain. A máháwat and his assistant, who had accompanied me and given unobtrusive advice, were to bring her, by easy marches, to my camp.

She eventually arrived and, for about four months, behaved herself admirably even proving staunch when facing a tiger. I was so pleased to see her steadiness, that I gave her the affectionate name of Guláb Pyári, which means Beloved Rose.

The day came, however, when she decided to reveal her true self by doing her very best to get rid of me and her máháwat. While on the march she suddenly began shaking her whole body, and continued to shake it in a most vicious manner. There was not a moment's pause, and the sensation was extremely unpleasant. The máháwat, gripping her neck with his legs as firmly as he could, speared her, and in his loudest tones advised me to jump down as best I could. All this time I was clinging to the pad ropes to prevent being thrown. The máháwat, being shaken all the while like an aspen leaf in the wind, cried: "This elephant means murder. Jump down and run."

Discretion was clearly the better part of valour and caring not how, I went. Partly by her tail, partly on my own, I slithered down and stumbled away, only to find that my ankle let me down. A few feet from me the quivering mountain of Beloved Rose thundered and plunged under the belabouring arm of her máháwat. I scrambled on all fours to a neighbouring tree and watched with relief the final victory of the máháwat's will, as he made her run to the stable. At the stable a conference of the máháwats recommended the use of a moogri as the only way to cure her of this new vice.

Henceforth a moogri formed part of Guláb Pyári's equipment. Shaped like a mallet with the head covered with three-inch nail spikes, it hung by a thin rope at the end of the pad, ready for instant use. The tables were now completely turned in our favour. At the slightest sign of a shake the máháwat would ask me to operate. Holding the moogri firmly I would bring it down forcibly with all its spikes on to her rump. She would immediately change her mind and run. How many laughs the máháwat and I had! Not until she had about a dozen lessons in good behaviour did Beloved Rose give up her shaking habit altogether. I unhesitatingly say that for a shaking elephant there is no corrective to compare with a moogri.

Not everybody realizes how absolutely indispensable to an elephant is its trunk. It is a deadly weapon, an extremely powerful long arm and hand combined, capable of immense strength in constricting, lifting, carrying, pushing, breaking, tugging, battering and throwing; it is a spoon for feeding, a pail to hold water, a hose, a sprayer, an indicator, a tester of stability; it is also a nose with a remarkably strong sense of smell. In short, the trunk is the elephant's very life. A trunk rendered useless through injury means a slow and certain death, unless special measures are taken to feed the elephant. In impending danger the trunk is often instinctively coiled up as much as possible and kept under the head in order to give it maximum protection.

I once saw a distressing incident in which an elephant's trunk was very severely mauled by a wounded tiger. The sportsman concerned did not shoot as accurately as he might have done. The tiger turned, charged one of the pad elephants and sprang on to its trunk, digging all its claws as deep in as it could. The distance between the máháwat's head and the tiger's was a matter of inches. In the scuffle the máháwat dropped his short driving-spear and could not recover it. There was the tiger clawing and biting the trunk. All the máháwat could do was to hit the tiger on the head with a stick.

The sportsmen, with their rifles ready, were all at impossible angles. To fire meant hitting the elephant or máháwat. It was not until one of them was able to reach right up, push the barrels of his gun against the tiger's side and fire that the beast released its hold and fell to the ground. The trunk was a ghastly sight. The flesh was hanging in long, thick shreds. The elephant was in great pain and the trunk hung useless.

She was under the care of a veterinary surgeon for about six months. She was fed day and night on sugar-cane, cut up in six-inch lengths, and pushed into her mouth by specially employed men, working in four-hour shifts. Water was poured down her mouth. The long, deep wounds healed in time, but the trunk was never the same, having lost inches in length. It was disfigured by deep scars all the way down, although the sensitive tip was fortunately unharmed.

At times elephants are puzzlingly temperamental and unpredictable. In the course of a beat planned for tigers lying up in dense, high grass, I was with a line of twelve elephants, all staunch to tiger, and moving along expectantly in steady, well-kept formation, when suddenly the whole twelve turned and ran as fast as they could. In spite of loud orders, mingled with the worst possible abuse, and much physical punishment from their máháwats, they had done the better part of fifty yards before they were halted and brought forward again under severe threats. And what was the cause of such a stampede and deliberate disobedience? The squealing of two tiny sloth-bears running along unseen in the grass in front of the elephants, left there by their mother while she went in search of food.

Compare the foregoing incident with this one. During a beat in which I took part, a tiger was wounded by one of the sportsmen. It went to a patch of grass about five feet high and sufficiently dense for the animal to be completely concealed. Among the elephants on which we were riding was a particularly black one with a very ugly face. As in the case of human beings, no two elephants are facially alike; the differences are readily noticeable to those who see them daily, as I have done. This elephant was exceptionally tall for a female; her blackness and ugliness made her conspicuous amongst her companions in the beat.

Hearing the tiger growling, and firmly making up her mind what to do next, she charged into the grass, unheeding her máháwat's injunctions to desist. Holding the tiger down with her trunk, she knelt down and rubbed him hard against the ground. At last her máháwat succeeded in making her withdraw. The tiger began to slink away but soon fell to one of the guns. We then had it brought out on to an open patch of ground to see its injuries. We found the spine so severely rubbed that the vertebrae could be easily counted; the skin was too badly disfigured to serve as a trophy. Even at this stage of the proceedings the máháwat was finding it difficult to keep the ugly one under control. Her intention was to attack the tiger again. She was indeed a fighter, with an inborn hatred for all tigers. In the scuffle she suffered several deep wounds from the tiger's claws and teeth. And this was one of the elephants which retreated helter-skelter from the squealing bear cubs no bigger than a fox-terrier.

In addition to the four elephants which were usually with me, two belonging to the Forest Service and the other two entrusted to my care by the Gun Carriage Factory, for whom

I used to do much honorary work concerning the selection of specially high quality timber in the Nepál forests, my friends and I obtained elephants from various large estate owners in the district. These were excellent animals, in perfect condition and accustomed to the hunting of dangerous game. In this manner the number of our elephants often varied between fifteen and twenty, which we found very useful when beating wide stretches of high and almost impenetrable grass.

Amongst these borrowed elephants was one named Gyankáli which I shall never forget. She was a dwarf, her height at the shoulders being not more than five feet. When I saw her for the first time with her máháwat, a much-wizened man with a scanty, grey goatee as his only protruding facial adornment, and whose stature fitted in admirably with Gyankáli's, I questioned her usefulness in a tiger hunt.

"You will see," chorused several máháwats who were by my side, and who knew the elephant well.

"What do you mean?" I inquired.

"We won't say anything now," came the reply. As if to make it more cryptic they added: "She doesn't take part in the actual beat."

Suspecting that the máháwats had a surprise in store for me I considered it tactful to ask no further questions.

When, next morning, the elephants were paraded for the day's hunting, and I saw tiny Gyankáli, with her little old máháwat sitting astride her neck and looking very important, I could not help but smile and go and pat her; I felt I could make a pet of her, and even permit her to enter the rest-house, to walk about the rooms and make herself comfortable wherever she wished. She could have walked through the doors quite easily, though there certainly would have been a problem had she wanted to lie down.

Everybody being present and everything ready, including the luncheon baskets and many soft drinks, for it was the height of summer on an Indian plain, we went to a portion of the forest where we had not been for a considerable time and where, therefore, the prospects of finding tigers were good. It was decided that the only animals to be fired at were tigers, leopards and bears. Sambur, swamp-deer, spotted-deer, barking-deer and pigs were to be ignored.

We searched for promising patches of grass. There was a relatively small one from which a deer, rising quickly, scurried away in safety. Perseveringly we continued and reached a large but gentle depression with a heavy crop of grass as high as the elephants themselves. Beginning at one end we proceeded to beat it slowly and methodically, with the elephants judiciously spaced to compel a tiger to walk ahead of us, and to emerge at the other end, where one of the guns on the flanks would have an opportunity of firing.

It was exceedingly heavy going as the grass was very dense, the stems thick, and of an unyielding quality. The elephants in the middle of the line had to force their way through with their very weight; they were completely submerged, the máháwats alone being visible; it was ideal cover for the animals we were seeking. On we laboured, when suddenly the sportsman on the left flank, a few paces outside the heavy grass, made frantic signs to tell us that a tiger had emerged some distance ahead and gone into the tree-covered area. Our quarry had succeeded in getting the better of us and despite our numbers we had clearly lost him.

But to our rescue came the diminutive Gyankáli. Called up from the rear, her máháwat was shown where the tiger had left the grass, and Gyankáli padded about like a hound. Then, the proud leader of a long line of elephants, she went ahead, with her máháwat uttering words of encouragement, and Gyankáli using her trunk in her own gifted way. The day's hunting had now taken a most interesting and thrilling turn. I was on the elephant immediately behind Gyankáli and so was able to watch her closely. Every few steps forward her máháwat said a few words to her; what they were I could not hear, and in any case they had their own language.

On the elephant behind me the impatient sportsman protested. "Where is she taking us? Surely the tiger cannot be lying up in such an open place under those trees."

I replied: "Let her go ahead; I want to see more of this intelligent trunk-work."

We went on for quite half-a-mile through the tree-forest. As we progressed through it, at Gyankáli's careful pace, the sportsman repeated that it was a waste of time, and others joined in with freely expressed opinions regarding the futility of going on and on, through these trees under which there was no grass for a tiger to take cover.

A solitary exception, I secretly shared all the máháwats' trust in Gyankáli. When I looked to them they nodded their heads, and made signs to me with their hands, to assure me all was well, and that the little elephant would lead us where we wished to go. In the meantime, the pessimism amongst the other sportsmen grew to such a pitch that they relaxed completely in their howdahs, and hope abandoned, placed their guns in the racks.

My eyes were focused on the working of Gyankáli's trunk, and particularly on its super-sensitive tip held steadily only an inch or two above the layer of dry leaves covering the floor of the forest. She was carefully scenting the leaves which were telling her a message she alone could interpret. Now and then glancing ahead, I saw signs of full sunlight, indicating the absence of trees, and, therefore, the presence of grass. Gyankáli was leading us straight there. It was a large patch of grass, sufficient in density and height for any tiger to seek cover in. Arriving at the edge of it she began to beat the ground with the end of her trunk when her máháwat immediately told her to stand still. There she stood triumphantly, as much as to say to all of us, and especially to the pessimists: "The tiger is in this plot of grass: I make no mistakes."

I at once alerted the relaxing sportsmen from their apparent coma. Signalling to everybody to follow me, and waving to Gyankáli's máhávat to take her to a place of safety, I went to the end of the grassy patch, formed the beat, placed sportsmen on the right and left flanks, telling them to keep some yards ahead of the beat. I then entered the grass at a snail's pace, in order not to cause the tiger to panic and gallop out and be lost again.

I knew from experience how tall grass was agitated in different ways by different animals. The movement caused by a bear shows a gentle, continuous disturbance. Deer and pigs, on being roused from their slumbers, rush out, causing an unbroken and distinctly violent agitation. What I now saw was a slight disturbance, then a tale-telling pause, followed by another gentle waving of the grass, then another pause, and so on. The object of the pause was to listen, evidence of the greater intelligence of tigers and leopards. I had no doubt that we were in the presence of a hidden but moving tiger trying to know what was happening behind him.

The tips of the closely-grown grass stems continued to be agitated in this sporadic manner. I signalled to the sportsmen on the flanks to be fully prepared, and pointed downwards to tell them what to expect in the course of the next four or five minutes. Slowly the trail of movement broke towards the edge, and then the tiger emerged, paused momentarily, and in that split second as he gathered himself to bound away, the sportsman on the left flank, before whom he had emerged, fired and dropped him.

"What do you now think of Gyankáli?" I asked, as we gathered admiringly round the dead tiger.

"Absolutely incredible," came the sportsmen's generous tribute.

Briskly and triumphantly Gyankáli was paraded by the proud máhávat; the sportsman whose shot had dropped the tiger stepped forward and placed in his hand a very generous currency note, at the same time requesting me to translate that half was meant for the máháwat, and half for Gyankáli in the shape of daily instalments of gur (solidified sugar-cane juice, pressed in the shape of large, rounded thick cakes).

Wonderful little Gyankáli! The máháwats, who had kept her secret so well, looked at me and smiled, emphasizing their smiles by fingering their greying beards with the lightest of upward touches. In my intense admiration of the tiny elephant I could find no better words than: "Niháyat umda" (most excellent), "shábásh, shábásh" (well done, well done).

There was a river known by the monosyllabic name of Ul (pronounced ool). When, standing on its high bank, I looked down upon it in the month of June, it was a mere trickle in a channel three feet wide and two feet deep. It crept towards the right abutment, wetting its foot and then snaked away in a dry bed stretching widely from bank to bank, spanned by a high bridge with many piers, three hundred feet long.

This bridge had borne wheeled traffic for more than half a century. For the inhabitants of the nearby town it was a vital link with many large villages from which essential supplies of grain and fresh foodstuffs of all kinds met the urban folk's daily needs.

Heavy and continuous rain fell in the month of August throughout the district. What was a mere trickle in June soon developed into a river covering the whole of the dry bed from bank to bank. The rain continued, the water rising rapidly till it was half-way up the high banks, and still the dark clouds hung overhead, pouring down blinding sheets of rain to drench and swamp the whole countryside.

The fast-flowing river rose higher and still higher, pouring urgently through the diminishing space between the bridge piers. As we made our daily, then hourly, inspection of the vital bridge a new and terrifying feature appeared. In the swirling muddy waters huge trees, complete with branches and roots came somersaulting, one after another, until there were scores in mid-stream, battering against the piers, closing the passages with their entanglements and thus forming a solid wall of flotsam, growing in mass and exerting a tremendous and relentless pressure against the whole structure.

To add to this crisis the abutments were being outflanked more and more as the river rose to an abnormal height. Looking anxiously at the bridge, and the great, green wall of trees banked up against it with no sign of the waters abating, we asked one another: "How much longer can the bridge last?" Not much longer. With a crack like thunder it was rent in three pieces, the substantial middle section being swept downstream as if it were a match-box.

Immediately the whole landscape was engulfed in turmoil. Men and women were carried away, clinging for dear life to the branches of trees, and crying out to be saved. I saw a large village hut being swept down with a family holding on to it. The river was now so wide that it extended far beyond its banks, and even threatened to engulf the town half-a-mile away. Whole villages, and age-old groves of mango trees, safe for many years, were uprooted and carried down. Hundreds of cattle, in the last stages of exhaustion, were endeavouring in vain to reach land. The entire scene was of an angry, relentless and cruel sea implacably dealing out death and devastation.

With no time to lose we commandeered every available elephant, and into this raging flood the valiant máháwats guided their indomitable charges. Suitably saddled for life-saving purposes, the elephants were swimming with the tips of their trunks a foot out of the water, the máháwats firmly straddling their necks, and naked except for a sparse loin cloth. Tips of trunks and heads of máháwats were all that could be seen in the merciless current.

Battling against floating obstacles as they came threateningly on, they worked in shifts, and in perfect unison. Back they would swim to land, each with its load of homeless village folk, to return again and again to the middle of the never-lessening deluge. These were deeds of outstanding valour, and as I witnessed them, hour after hour, I found in my heart a new respect for these truly noble animals and for their skilful and courageous máháwats.

CHAPTER VI

THE MAN-EATER OF PAHÁRPANI

DURING MY CAREER in the Indian Forest Service I spent no less than fourteen years in Kumaon, the land of man-eating tigers. In spite of such hungry, roving and terrifying carnivora, there was for me no more delightful country in all the wide world.

I traversed every mountainside. The climate was exceptionally bracing. And wherever I went I found the people friendly and co-operative.

My tours occupied nine months in the year. Only in July, August and September was camping impossible on account of heavy, continual showers of rain, bringing the annual rainfall up to a hundred inches. Though snow fell during December and January it did not detract from the pleasures of an outdoor life.

It was during one of such tours that I met the man-eater of Pahárpani, a camping site about 6000 feet above sea level and situated at the head of the Gaula valley amongst a forest of oaks and rhododendrons.

Work in this valley kept me extremely busy. In addition to settlement operations there was much to be done in connection with the commercial exploitation of the pine forests. To aggravate matters, three of my tree-marking officers suffered from malaria and were bed-ridden. The situation was developing into a crisis as far as I was concerned; so much so, that I decided on doing tree-marking myself.

While I was thus engaged one of my men called out to me in urgent warning. He was higher up the hill, while I was studying the silvicultural aspect of certain trees about twenty-five yards from him. I was unable to make out what he said but got the gist of it sufficiently to stand alert with my eyes on the intervening ground. Almost immediately I saw, some eight feet to my left, a golden-coloured cobra. Erect and with hood expanded it slithered straight down the slippery mat of amber-hued pine needles. There were no stones at hand so I picked up a small branch and hurled it as the snake approached. In doing so I slipped and fell and for a moment felt true panic, but the cobra continued its precipitous journey downhill.

That same day crossing the Gaula river while returning to my camp at Khansyun, I saw two very black cobras, the exact colour of the rocks in the bed of the stream. I was then in a better state of mind to admire these perfect cases of Nature's camouflage.

A day or two later came disturbing news from the Range Officer of the territory concerned, that building work on the rest-house at Pahárpani was held up by the presence of a man-eating tiger.

Pahárpani was at the extreme head of the valley; it was a road junction and so a convenient site for a rest-house. The plan of the new bungalow provided for a verandah, two main bed-sitting-dining rooms and two bathrooms, but these features meant less to me than the requirement that the construction must be completed by the end of March, the close of the financial year.

Harlál, the contractor engaged in constructing the house, reported that it was impossible to complete the project in time; his masons and carpenters were deserting him in their fear of the man-eater which had recently killed a villager very near Pahárpani.

This information was doubly unpleasant. If the rest-house was not completed in time, and payment made to Harlál, it meant that the substantial sum of money allotted for the purpose would lapse. In the world of officialdom such lapsing of funds was almost tantamount to a criminal offence. It was argued that an officer had no right to include in his budget an amount which he could not spend within the financial year. To make excuses was futile. They were never accepted by very important persons who, as they ascended to the highest rungs of the ladder, were prone to lose complete touch with the difficulties of the executive branch of the administration, difficulties which included such matters as malaria and man-eaters.

There was only one decision in such circumstances . . . to go to Pahárpani and, on arrival there, to stand over Harlál's workers with a gun. This, I hoped, would instil them with a sense of urgency as well as confidence, and so enable progress to be made with the building. I therefore altered the whole of my itinerary.

Moving camp daily, I reached Raikuna and spent the night there. The following day I was on the march to Pahárpani. I was within two hundred yards of it when, as was my practice, I looked down from the right edge of the road to a small plot of cultivation in a depression with a tiny hut as a humble homestead. I expected to exchange greetings with Gusain, its owner. He usually came up to the road to meet me. But on this occasion there was no attempt to do so. He sat at the foot of a tree, a few feet in front of his hut, and ignoring my call he continued to sit with his head buried in his hands.

Gusain belonged to the caste known as dom, and he was actually dubbed as an untouchable. He could only give his sons and daughters in marriage to the sons and daughters of another dom, and to no other caste. His one comfort was he was a Hindu; but even so he could not enter a temple.

Watching him for a few moments, I concluded he was not well, so I went down to see what was wrong, and to know if I could do anything to help him and his family. As I approached I saw that his lean, ill-nourished body was shaken with deep sobs. I could see nobody else, the hut stood deserted, and there were none of the usual children running around.

Eager to know the cause of his sorrow, I waited for him to recover. When he had done so slightly, he looked up at me, the tears still streaming down his cheeks. He could scarcely speak, and I found it difficult to make out his few incoherent words. I sat on the ground near him and gently asked him to repeat what he had said. Speaking haltingly through his sobs, he said: "He has taken my wife and children, now I am all alone. A few days ago he came again. I had a very narrow escape. I climbed this tree just in time, but it would have been better had he taken me too."

Gusain did not mention the word for tiger, but referred to the animal as "wuh" which, translated, is "he" or "it". There was no need for me to ask him what he meant by "wuh". I had already understood. All I could do to comfort the old man was to give him money to buy a generous quantity of grain and stores.

While he remained silent wiping away his tears, only for more to cover his face, I studied the environments. He had about an acre of arable land; the meagre crop on it was in urgent need of weeding; it looked so neglected. Speaking again, Gusain said: "This is my home; it is all I have; where can I go?" I felt for him very, very deeply. His loss was irreparable. I said I would see what could be done for him, adding I would do my utmost to kill the man-eater.

The surroundings of Gusain's modest holding were very much in favour of the man-eater. There was an oak forest with dense bush, giving the tiger excellent cover and enabling him to approach near the hut; he could spring on his victim, giving him or her no chance whatsoever. Gusain had built himself a platform in a tree, between six and seven feet from the ground, and from this he had a better view of the surrounding undergrowth. I gave him permission to cut away the bushes as far as he liked in order to improve visibility all round this arboreal perch. When he was able to talk a little more I inquired when the man-eater had taken his wife and children. "Over two months ago," he replied. Then with a clearer remembrance of the time, he added: "Before the contractor and his men came to build the bungalow."

Leaving Gusain, I went to my tents which had been pitched near the site of the partly completed rest-house. Harlál and his masons and carpenters expressed themselves pleased to see me. They were all huddled together in their temporary hut, and like all hillmen when not working, they were smoking their chilums (small conical-shaped clay pipes held in both hands while inhaling). No time was lost in discussing the situation. I assured them of full protection against the man-eater at least during daylight. I pointed out, however, that much time had been lost, and inquired what they were going to do about it.

I was pleasantly surprised when, with one voice, they said that now I was with them they would work overtime, do their utmost to complete the building in the shortest possible time, and certainly before the end of the financial year. I suspected that the contractor had offered them handsome rewards. He could indeed afford to do so as the rates for all items were generous. But to see the men's spirit of co-operation gave me great comfort.

Answering my inquiries about the tiger's latest activities, they could give me no definite details beyond what they had heard of a tragedy between Pahárpani and the next camping place, Mornaula. This latest killing, and those that preceded it, indicated that the tiger was a growing menace to this particular locality.

Standing near the unfinished building, I decided where to station myself while the masons and carpenters were at work. Below the site was the main road from Dhári; here the bridle-path from Raikuna met it. From this junction the main road continued on to Mornaula, about five miles distant in a general easterly direction. Visibility was good in all these directions, and it was most unlikely that the man-eater would expose himself by walking along these roads during the daytime.

The danger lay to the north where there was an oak forest which came down almost to the site itself. Under the trees there was enough undergrowth to give the tiger all the cover he required to make an unseen approach. The width of the clearing between the site and the scrub was not more than twelve yards. Understandingly, these conditions had made Harlál and his employees nervous. Being unarmed, no one could possibly blame them for refusing to work; it was just fear, and not a strike. The tiger could have charged out of the undergrowth and sprung upon any of them while they were fashioning stones or making door-frames. Here everything was in the enemy's favour. Accordingly I decided to stand, with my gun, between the unfinished building and the forest.

Punctually to the minute, the workmen resumed work the following day. For my part, I loaded my double-barrel rifle and stood, vigilant and ready to shoot, looking keenly to right, left and centre for a glimpse of a long, tawny, black-striped body. This guard duty lasted with brief intervals for meals for at least eight hours a day, and sometimes longer; but it had to be done if the building was to be completed in good time.

In such circumstances, there arises in one's mind the fullest appreciation of the vast difference between stalking and being stalked. As you stand, and continue to stand, hour after hour, and day after day, the first resolute feelings begin gradually to disappear, some measure of relaxation begins to set in, and, unless you keep yourself up to the required degree of vigilance, extremely peculiar feelings begin to run up and down your spine. "Very definitely," I used to say to myself, "I prefer to stalk a man-eater than be stalked by him."

No incident occurred. Daily I stood like a sentry, but with the considerable difference that whereas my hours were very much longer, the alertness demanded was no less than that required in war. All the time I was hoping most fervently that I would receive news of the tiger's whereabouts. As a Forest Officer, my position was very different from that of a free-lance hunter. The latter's time is his own; he is at liberty to spend week after week in pursuit of his quarry. A Forest Officer, on the other hand, had to account fully for his time and actions in a diary, or a fortnightly report, submitted to his immediate superior. Such a document containing a sentence: "For the past two weeks I have been watching for a man-eating tiger," would most certainly have been promptly returned with the telling comment: "Such a pastime should be left to those who have unlimited leisure at their disposal. Your inspections have undoubtedly suffered."

Without fear of such an adverse remark, the longest time I could spend at Pahárpani was ten days, with a full explanation justifying such a maximum period at one camp. I decided to make the best use of the time by sending out messages to the various headmen of villages to send me information about the man-eater's activities.

The first news I received concerned a boy, about fifteen years of age, who was sent by his father to bring his pony back from grazing to enable him to stable it for the night. The animal had wandered along the road and then decided to graze in a glade surrounded on three sides by an oak forest, and on the fourth adjoining the main Dhári–Pahárpani road, but nearer Pahárpani than Dhári. The boy found the pony; picking up a stick, he was determined to enjoy a gallop homewards. In his exuberance he mounted at one end of the glade, used his whip freely and his steed obeyed. At a gallop he was approaching the other end of the glade, in full view of two wayfarers, when the pony swerved at almost right angles, throwing the boy on to the ground.

In a second the tiger, crouching on the edge of the forest and waiting, pounced on the unfortunate lad and carried him away. The pony had suddenly seen the man-eater; the boy in his excitement had not. To its home the pony went as fast as it could. The two wayfarers returned and gave the sad news.

It was too late for a beat. Early the following day we searched the whole forest and beat it, but there was no sign of the tiger or the remains of the boy. Three days later the headman of Raikuna arrived to tell me that the tiger had been seen above his village in the forest about two hundred yards above the houses. There was a definite reason for the man-eater to be wandering about there. In Kumaon fresh oak leaves are gathered as a favourite fodder. While the men work in the fields, the women, after completing the morning's chores, take their dharántis (heavy sickles made for cutting branches) and go in groups to the nearest oak forest to lop the trees. Each woman returns with a large head-load of leaf fodder.

A man-eating tiger of this district usually is well aware of this practice, and he fits in his movements accordingly. His adopted way of life compels him to follow the ways of man; so he will wander by day and sleep by night. He will, temporarily, reverse this order if his appetite has been satisfied, but otherwise, his hunting is done by day. That is why he was seen above Raikuna village. He was waiting for a group of women to come up, as usual, for leaf-fodder, intending to snatch one of them on the ground, or from an accessible branch of a tree. During my years of travel in Kumaon I heard of several cases where women had been dragged down, from low-branching trees, by man-eaters.

At nearly all road junctions in Kumaon there was a small shop where the owner did a moderate amount of business with wayfarers who would sit on a bench, have a snack, a smoke, and impart the latest news. The shopkeeper as a good trader would pass on the important items to others. Pahárpani boasted such a gossip shop. The owner prospered, and, in time, employed a servant whose wife went daily to the nearby oak trees and brought back leaf-fodder for the shopkeeper's cow.

While she was busy in the tree lopping the small branches with their leaves, she felt the tree violently shaken. Looking down she was horrified to see the man-eater. He had struck, hoping to reach her left leg which, although on a lower branch, was luckily out of his reach.

Instinctively making her position safer by climbing higher in the tree, she called as loudly as she could for help. Her husband, who was busy at the shop, answered her and then ran to me. I was standing guard over the masons and carpenters who had also heard the woman calling out, and had stopped work to listen intently. Advising them to retire to their hut and to remain there, I went in search of the man-eater. There was none sufficiently reliable to accompany me with a second rifle. I, therefore, desired no company.

My immediate problem was to find the precise whereabouts of the tiger. Was he still close by? Or had he wandered off through the forest after he had failed to pull the woman down from the tree? My only source of information on these important questions was the woman in the tree.

Moving warily along the road I looked above and below it, expecting a charge and a spring at any moment. At the same time I glanced up into the trees to catch a sight of the woman. Three or four steps further on I saw her. She had evidently been watching my progress, as our eyes met at once. I made a sign of inquiry. She pointed down to her right. This I took to indicate that he had not wandered far from the scene of his attempt, but still the question in my mind was: where precisely was he? It was essential that I should see him first, and not vice versa.

Tiptoeing four or five steps further, I looked up again quickly at the woman. The tree on which she was now safe, was only a few feet from the lower edge of the road. A few yards further on was a dry watercourse above five feet wide and four feet deep. From the tree to the watercourse, in a south-easterly direction, I judged the distance to be eight yards.

In response to my second inquiring look, the woman pointed again to the same spot. Straining my eyes in that direction I still could get no glimpse of the tiger. Standing motionless for a second or two, I decided that the only place he could be in was the watercourse, the head of which was just below the road. I had passed it during my inspections and noted it was a danger to the stability of the lower edge of the road as erosion increased year after year.

I now felt the need of a companion who could have reconnoitred this tricky bit of country but on my own there was only one line of action: that was to approach the head of the watercourse by continuing along the road and looking down it. The question now was: which of us would see the other first? Putting my rifle up to my shoulder, since every second now counted, I stepped forward to obtain a quick view of the watercourse. There was the man-eater, facing upwards, towards the woman.

The moment he saw me he snarled fiercely and I was aware of him coming straight at me, but simultaneously my first bullet found its mark and he fell. As he did so, I fired again and reloaded. Keeping my aim on him I watched for a moment. He was still. It was the end of the man-eater of Pahárpani.

I asked the woman to call to her husband to bring several men and a chárpoy (Indian bed), since this would be the most convenient and the quickest arrangement to carry the tiger to the camp. The men came running along as though they were having a race. It took their combined efforts to drag the man-eater out of the watercourse, and place him on a chárpoy to carry to the camp.

He was well past middle-age. His canine teeth were intact but he had lost all the incisors. Two molars were missing. On examining his legs I found two claws broken. Otherwise, he was in quite good condition. From his nose to the tip of his tail he measured nine feet five inches.

As usual, there were many requests for the man-eater's fat. The people in the hills, like those in the plains, believe firmly in its efficacy for rheumatic complaints. The shop-keeper, Tára Dutt, a Kumanoi Brahman, was so jubilant at the death of the man-eater that he served out gram and jaggery free of cost to all present. Knowing his reputation for stinginess, I looked at him in surprise. "It is quite all right," he remarked, "there will now be more wayfarers on this route and, therefore, I will do more business."

That very lonely figure, Gusain, had left his hut and now sat staring at the tiger that had taken his wife and children. The tears ran down his cheeks when I spoke to him. No act of mine could bring a full measure of comfort to his deeply-scarred heart. The most I could do, to bring some comfort into his shattered life, was to appoint him, on a monthly salary, as caretaker of the rest-house as soon as Harlál, the contractor, had completed it.

Twenty-three years later, I revisited Pahárpani, again in an official capacity. The scene had changed. Gusain had passed away, so had Tára Dutt. There was not a face I could recognize. The oak forest and the scrub, where I had stood like a sentry day after day, had been completely cleared, and in their place was a large field covered by a flourishing crop of potatoes.

In all forest rest-houses there is a visitors' book. When leaving Pahárpani for the last time I recorded in it, for the information of future generations of Forest Officers, the somewhat unusual circumstances in which the construction of the rest-house had been completed on schedule.

Before finally leaving this area I would like to pose a question of unusual interest: WHY ARE THERE MAN-EATING TIGERS ALWAYS IN KUMAON? I have read every book published on man-eaters in this tract but nowhere have I seen this question discussed and answered.

There has never been a time when this territory has been without man-eaters. I heard of them when I first arrived in Kumaon in the year 1912. I have heard and read of them ever since, right up to the end of December 1955, when I was still in India. That means a period of forty-three years, at the end of which the Government were continuing to offer rewards for the destruction of man-eaters, and were even dispatching small parties of infantry and police to deal with the menace.

As I see it, the reasons for a tiger turning man-eater in Kumaon are:

(*a*) Scarcity, or complete absence, of his normal food which consists of all species of deer, pig, cows, bullocks and buffaloes. This is the principal reason.

(*b*) One or more disabilities brought about by wounds caused by:

- (i) inaccurate shooting or poor quality guns with inferior or unsuitable ammunition;
- (ii) mishaps on unfamiliar terrain, such as steep hill-sides;
- (iii) suppurating injuries due to the penetration of thorns and porcupine quills, or from fights with other tigers.

(*c*) The increasing handicaps inherent in advancing years, such as reduced agility with consequent inability to seize fleet-footed herbivora on mountain-sides, and loss of its weapons, the claws and teeth.

If all the above reasons are carefully considered it will be seen that they can be summarized in one word—HUNGER. There is the additional possibility of a man-eating tigress feeding her cubs on human flesh and so bringing them up to man-eating. This cannot be ruled out as such cubs are bound to kill men, women and children, if they continue to live in an area of scarcity. Again, it is a question of hunger. There comes a time in the lives of all tiger cubs when their mothers no longer wish to have anything to do with them; they are made to understand that they must go and fend for themselves; if they persist in staying, then the mothers use violence, especially if about to have cubs again.

We must, however, go back in their life-histories in order to obtain a comprehensive understanding of how and why these man-eaters enter Kumaon, and will continue entering a territory that will always be characterized by an acute and growing scarcity, in very many places a complete absence, of a tiger's normal food.

Kumaon comprises the two districts of Naini Tal and Almora. Their southernmost parts are a long, unbroken range of mountains known as "the foothills of the Himalayas". These give way in a southerly direction to a wide stretch of gently sloping ground which continues till it terminates in flat country extending away into the plains. Thus there are three distinct zones. It is the middle one, the terrain with the gentle slope which is so important. It is richly forested, holds an abundant stock of deer and pig and has been, from time immemorial, an ideal breeding-ground for tigers. It is essential to add that here they do not molest man, unless molested by him.

The dense forests continue for a portion of the way up the foothills, merging with increasing altitude into the oak-pine flora, so that a tiger can enjoy continuous and sufficient cover when travelling from the middle to the upper zone. But why should a tiger quit the zone which is so well stocked with his normal food? The answer is the Darwinian phrase: *the struggle for existence, and the survival of the fittest.* It is a fact well known to Forest Officers that non-man-eating tigers and tigresses are local and, as a general rule, pairs keep their own well-defined territory. In due course the pair have a family, usually of two cubs. As these grow up they develop sufficient strength to become, in time, a match for their parents and become severely intolerant of their presence with the result that the older pair will be driven out of their beat. Occasionally the youngsters go into the adjoining beat and drive out the older generation there. This expulsion or destruction of the older is, of course, nothing new, and has been going on in the jungle from the beginning of time. The aged, increasingly enfeebled, must give way to the young, growing in strength.

Losing the battle, or avoiding it, the expelled tiger seeks refuge in the adjoining hilly zone of Kumaon. In doing so he unknowingly enters a strange tract where life is extremely hard, his normal food is too scarce to sustain him and the circumstances, taken collectively, so thoroughly unfavourable that he is compelled to choose a new way of life; the pangs of hunger sting more and more and point in the inevitable direction of the easiest of prey—human beings.

In other words, in accordance with an inexorable law of Nature, the younger generation of tigers in the breeding-grounds are constantly pushing the older into the hills where a tiger finds it impossible to live unless he resorts to human flesh.

This, then, is the reason why there are, and always will be, man-eaters in Kumaon. No matter how many may be killed from time to time, others will appear in their place.

Tigers, which have entered the foothills, go occasionally further into the mountains. Such journeys are the result of the annual movements of villagers with their small herds of cattle. When they return in spring from the sub-montane tract, their herds are sometimes followed by tigers hoping to get a straggling cow or bullock, or one which lingers by the wayside through sickness or weakness. That is why a man-eater is encountered in higher altitudes, like Rámgarh, Muktesar and Ágar, all of which places are from 6,500 feet to over 7,000 feet above sea level; he continues to follow the herds which, with their owners, eventually scatter to their respective villages, leaving the tiger stranded. In the new locality he must, in order to keep himself alive, continue to kill and eat human beings.

CHAPTER VII

PHOTOGRAPHING WILD ANIMALS

THE VERNAY-FAUNTHORPE EXPEDITION

IN THE HOTTEST PART of the summer, in the year 1923, I received orders from the Government of the United Provinces to give all possible help to the Vernay-Faunthorpe expedition whose object was to obtain the best available specimens of wild life for the New York Natural History Museum.

Arthur Vernay, the leader of the expedition, was a sportsman of the highest calibre. So particular was he about the subject of game preservation that he would never press a trigger unless the animal was a really good specimen worthy of exhibition, and in the best interests of Natural History. There were many still to be collected for the Museum, and it was Vernay's aim to supply as many as possible by making extensive tours in the East. He had made up his mind, unselfishly, to devote part of his wealth to the advancement of Natural History.

Mrs. Vernay accompanied him, and I have yet to meet a more charming and considerate hostess. Wherever the Vernays travelled in their quest they left the most pleasant of memories behind them. It was a real pleasure to assist them in every possible way to achieve their ambition.

Faunthorpe, of Bisley fame, can best be introduced in this episode as one of the late Lord Morley's "sun-dried bureaucrats", and a member of that body of Indian Civil Servants whom the late Earl Lloyd George described as "the steel frame" of the British administration in India. It was towards the end of his long service in India that Faunthorpe decided to join Vernay, and in doing so helped to make the expedition the great success that it was. He was then Commissioner of the Lucknow civil division comprising several districts.

Tall, chisel-featured and broad-framed, Faunthorpe had a commanding personality. In addition to his official duties, he gave part of his time to the Auxiliary Force of India and was colonel of one of the battalions.

A rural existence was his very soul, the forests and their fauna his paradise. He had no use for such institutions as a Secretariat, for which he expressed a profound contempt. As a junior Civil Servant he rejoiced in pig-sticking, his horsemanship being of a high order, and his supreme feat was when, on horseback, he chased a leopard and in a fierce scuffle speared it.

As a rifle-shot, whether at a stationary target, a fleeing deer, or a vanishing tiger, he had no equal. When he was nearing fifty years of age his large blue eyes were as keen as in his youth; every shot at five hundred yards on a rifle range was a bull's eye. We often hunted together; never did I see him miss. A singularly striking feature about him was his hands. He would relate a shooting incident in his slow, deliberate manner, at the same time placing his right hand over his left arm and gradually work it upwards to twirl his moustaches. All the way it would tremble severely. I used to wonder how such a shaking hand could ever hold a gun steady. Remarkable to record that when a gun was in his hands they were the steadiest you could ever wish to see. Faunthorpe was the firmest of friends. Our association in the forests during several years has left the happiest of memories.

Vernay completed the personnel of the expedition by employing a cinematographer named Dyott. It was a good plan to make a comprehensive record of the work done by the expedition, and to complete the record by taking as many scenes as possible of animals in their natural haunts.

Another, and a very essential, person employed by Vernay was John Jonas, taxidermist, an Austrian by birth, but an

American by nationality. His skill was exceptional. His mountings (Jonas vehemently objected to the term "stuffing") of animals in the New York Natural History Museum were works of outstanding art.

Jonas's work in the forest was only the first step in his painstaking operations. When an animal, such as a tiger, was shot, it would be carried on the back of an elephant to camp and deposited under a distant tree, sufficiently distant for the intervening space to act as an efficient buffer against unpleasant odours being wafted to our tents, to enable him to work in solitude and so devote the fullest degree of concentration to his tasks.

His outfit included several registers, on the pages of which were diagrams of animals with each important part designated by a letter or a number, and with a reference column for the letters or numbers. Fine taxidermy demanded the fullest details on the spot, without which it was impossible to achieve a faithful mounting while working in the Museum.

As the skinned tiger lay on its side before Jonas, he would pierce the shoulder with a long, stout, sharply-pointed steel needle down to the bone. He would then withdraw the needle and measure meticulously the degree of penetration; this he would enter against the letter "A". In this painstaking manner he would record the detail concerning each part of the body. These measurements were to be his guides in effecting the natural curves so very important during mounting in the museum. To him his completed diagram was a priceless record.

But, of course, mounting in a museum cannot be done without bones. Accordingly every bone in the tiger had to be separated and labelled. For this purpose all flesh had to be scraped off and the naked bones thoroughly washed and antiseptically treated, the skull and adjoining parts received special attention. After treatment all the bones of one animal were placed in a separate package together with the cured skin, and dispatched to New York at once, so reducing the expedition's packages to the very minimum.

I marvelled at Jonas's nose. It was truly extraordinary in that it was definitely proof against the strongest stench. He would work assiduously and continuously at a tiger's carcass, several hours old, in the hottest of hot summer days. It would be stinking to high heaven. Yet Jonas, without a handkerchief across his nose, would bear it all without even a grimace. I used to stand very many yards away as an onlooker. Professional skinners of carcasses would walk away and expectorate at intervals. There was no internal for Jonas, and no expectoration. Buried in the awful stench, with his head well down over the carcass, he would continue his work unaffected by the overpowering odour. The professional skinners, known as chamárs, with whom high-caste Hindus would never deign to sit down to a meal, since such unfortunate humans were viewed as untouchables and even forbidden to enter temples, would shake their heads in wonder at the stench-proof Jonas who, in spite of all, never suffered a day's illness. His was an achievement in immunity.

One evening, after the day's hunting, Faunthorpe stepped into my tent and, as he often did, flopped down on the bed. Lighting a cigarette we discussed the day's doings and eventually the subject of photographing more scenes with wild life in action received our attention. Vernay made a timely appearance. The evening's discussion was therefore a full one. The outcome was the decision that Faunthorpe and Vernay would continue their efforts to obtain specimens in the dense forest, while I would take Dyott to the open country and obtain pictures of spotted-deer, black buck and blue-bull. These were the three species usually found there. The tract was flat with a narrow river flowing through it, the trees were scattered and there was good pasture. Legally, the area was a reserved forest set aside for the grazing of villagers' cattle. I was to be given a dozen elephants and as many men as I considered necessary to drive the animals to Dyott. I lost no time in summoning all concerned and issued instructions for the next day's operations.

Breakfasting early, Dyott and I packed all necessary equipment into a model T Ford and went to the scene of our activities. We took the car as far as it could go, and from there had the equipment carried carefully to a central spot. It was towards the end of May, which meant that an extremely hot day's work lay ahead of us.

A problem immediately presented itself. The camera Dyott was using operated on a tripod. We therefore had to search for a tree with spreading and almost horizontal branches on which the tripod could be fastened, and construct near it a Heath Robinson type of platform on which he could stand. This search occupied a considerable time. We examined many trees and, after nearly an hour, succeeded in finding one which appeared to meet our purpose. But the branches of the tree spread in such a manner that though the camera could revolve on the tripod in every direction, the operator remained stationary on the platform; this restriction of movement meant that the animals had to be driven up to Dyott within a limited arc, thus making my task all the more difficult. Fixing the tripod, with its camera, to the branches and fastening the plank platform, we helped Dyott to ascend and take up his position.

When he gave the sign that he was ready, I departed on an elephant accompanied by two others on either side of me, and spaced out about thirty yards, in search of a herd of deer. Moving cautiously and very slowly, looking in every direction, we were at last rewarded when we caught sight of a fairly large herd of spotted-deer.

Making a wide detour, in order not to upset the herd prematurely, I arranged the formation in a rough semicircle, with the elephants well spaced, and the men spread out behind them to prevent a break-through. We now had the herd between us and the camera. I gave the signal to advance at the slowest possible pace, having already instructucted everybody to keep their eyes on me for signals, as there was to be no talking. As the elephants advanced the spotted deer looked up from their grazing. The herd numbered between twenty and thirty, there being only four stags of various sizes. I continued to advance on them very gradually to induce them to go forward at a walking pace. Not unduly disturbed, they walked, trotted slowly, then stopped to look back at the elephants. Now and then I halted the elephants to give the herd time to gain confidence, and to conclude that we were not so dangerous as we appeared to be. This had a good effect and the deer seemed more trusting. We were still quite a quarter of a mile from Dyott. Many and frequent were the signs I had to make to the máháwats to go slower, or to halt. I realized that if a single elephant went forward beyond the set pace the whole herd would stampede, and it was the stampeding that had to be prevented at all costs; it was in fact the very essence of the whole operation.

Dyott now caught sight of the spotted-deer approaching him. I had already warned him not to move about on his platform lest the herd should see him and run. It must have been very tiring for him to stand on those planks waiting for something to happen, but not so wearing as my anxiety that the herd should be controlled, and eventually walk in front of him.

The climax was not long in coming. The spotted-deer went along at an easy pace, walking, then trotting, and walking again. In this way they progressed till they were not more than twenty yards straight in front of Dyott, who was actively working the camera. And then, to my surprise, the whole herd stood still, in response, it appeared, to the elephants already being halted. There they stood, shaking their heads and flapping their ears to keep the flies away. It was an extremely satisfactory culmination to all the manœuvring that had been done. Dyott and I were delighted.

From my knowledge of the three species concerned—spotted-deer, black-buck and nilgai—I knew only too well that I had dealt with the easiest of the three, and that my hardest task had yet to be tackled. As the elephants came up to the camera the spotted-deer disappeared at a fast trot, the tempo of which quickly rose as the leader quickened her pace to a gallop.

After a brief respite, during which the máháwats and the men on foot ate gram, and the elephants blew saliva over their bodies with their trunks, I arranged that I would go in search of nilgai and black-buck. As Dyott had never seen these two species I described them to him, taking care to emphasize that the former were almost as tall as horses, and that the bull was a dark slate-blue in colour.

Riding an elephant, and taking two more in extended order, I scoured the tract. My effort was rewarded when I saw a nilgau (bull) and four nilgai (cows) standing under a heavy-crowned tree enjoying a noonday rest. But they saw my three elephants as quickly as I saw them. They began to fidget and I withdrew at once, directing one of the máháwats to retreat and conceal his elephant in such a manner that he would still be able to watch the five animals. I next returned to the rendezvous, collected the remaining elephants and all the men and decided on a very wide detour to get the bull and his four cows within a semi-circle. I explained my plan to Dyott and warned him about the direction from which the five animals would come. The scout I left behind came and informed me that they were still under the same tree.

Forming an arc and then closing in on the flanks to make the general formation approximate a semi-circle, I adopted the same tactics as with the spotted-deer. But the blue-bull was inclined to gallop, the cows following him. I was obliged to halt the elephants again and again. So long as the five of them were within the semi-circle I felt satisfied. At times it became necessary to signal the extreme flanks to go ahead, to keep the blue-bull and cows just where I wanted them, and not to allow them to break out of the formation. It was interesting to see the bull amble now to the left, then to the right, as he saw the elephants in both directions. He had not broken into a gallop, and, so long as he trotted, all was well. I knew that the moment he galloped it meant a stampede, and Dyott would not get a chance of taking a picture. The cows obediently followed the bull. Their present formation and rate of progress was entirely satisfactory.

Now we came in sight of Dyott, and by his movements I could see that he was vigilant and fully prepared. It was now a matter of rapid hand signals, telling the right flank to move ahead, then instructing the left to do the same. On trotted the bull and the cows, like horses loose in a paddock, turning their heads all ways. I became distinctly optimistic and, later, delighted, when I saw the bull and his cows approach Dyott and trot past him not more than twenty yards from the camera. I felt I had succeeded with the most difficult of the three species. And I felt sure that some rare and interesting pictures had been taken.

I had already halted the entire line to give the bull and his companions confidence. A few paces after they had passed Dyott something scared them and away they all went at full gallop. Giving the signal to advance, we went to Dyott's tree to hear his comments on the success of the whole manœuvre.

Great was my surprise and disappointment when he descended from the tree very red in the face, and said with much feeling that he had not taken a single picture. His gestures had been an attempt to let us know that his camera had completely jammed. In the keenness of our disappointment we each felt very disgruntled with the other.

We packed up the camera and I indicated another elephant for him to ride on and reach the car. Just then I wanted none of his company. We reached the camp in silence. I went to my tent to have a cup of tea, and to see what the daily post had brought in the way of official and private letters. I was busy reading these when Faunthorpe and Vernay returned from their daily excursion to the forest, to hear of our mixed bag of success and failure.

The following day I was unable to join the photographing unit as I had already fixed the date for settling a case of forbidden romance in which two Tháru families were involved. To postpone a settlement would have been unkind as well as unjust, especially since it was a serious matter from the two families' point of view. So Dyott spent the day in developing some of his negatives.

The Thárus were my people. I was directly responsible for their welfare. They occupied forty villages on the borders of the forests, but the land on which stood their homesteads as well as the plots they cultivated, came within my jurisdiction.

Because of their history and traditions they were a singularly interesting people. For their livelihood they depended to some extent on agriculture, but for about six months in the year they found employment with timber contractors. A river, about five hundred yards away from their hamlets, supplied them with an ample diet of fish.

The Thárus were a brown-complexioned race, clean in their habits, of average stature, the men well-proportioned and muscular. They tucked their longish hair in their small, carelessly-tied turbans, but the women were very dress-conscious, and wore their hair high on their heads, Spanish style, their saris draped in graceful spiral folds from the top of the coiffure to their ankles. The colour most fancied was a dull red. To them silver jewellery was irresistible; necklaces weighted with bright red beads were always conspicuous. Their homes were spotless. Pride of race precluded the women from seeking outside employment. Their queenly bearing and manners pointed to a link with Rájputáná, though this particular territory was many hundreds of miles away.

A striking feature about the Thárus was their relative immunity from malaria. They were the only people who lived in this particular forest zone all the year round, and were able to survive the notoriously unhealthy rainy season, from the last week in June to the middle of October. During this period the entire tract was infested with the Anopheles mosquito which bred here in its myriads; so much so that the entire forest staff were given special leave in those months. I made a few excursions to learn exactly what were the conditions in the forests during that period. With a luncheon basket I would sit on an open spot, such as a wide fire-break. In a few seconds, there would be humming around me, hungering for my blood, scores upon scores of mosquitoes. Those which rested on my hands and legs did so in the semi-vertical, head-down-tail-up attitude, so characteristic of Anopheles. It was quite impossible to finish my lunch in such circumstances, and I was forced to pack and quit.

On the early history of Thárus, and whence they migrated, historians have not been able to agree, and I ignored their theories and adopted the much more fascinating and direct method of sitting among their elders who discoursed at length on their origin and subsequent movements. It has been a custom, almost as old as the Earth itself, for all illiterate tribal folk to pass on their own particular lore from generation to generation. Their memories are their archives, always engrossingly interesting, even after making due allowance for a pardonable measure of embellishment.

There was one particular custom amongst them that I singled out as unique, one that I am not aware exists anywhere else in the world; the wife cooks the food, eats her share in an inner room and then hands out her husband's to him, to be eaten in an outer room. When I told them that I had never heard of such a practice one of the senior dames drew herself up proudly and explained: "It is a matter of tradition. In bygone times there were many inter-clan wars in our country, Rájputáná. When a clan was fearful that, in the event of defeat and the deaths of all their soldiers, their wives would be taken by the victors, they took the precaution of sending their womenfolk to distant places with their male servants in charge of them. Then it happened that our clan lost a battle, and the wives never saw their husbands again. The women were thus left with their male servants. Marriages of a sort between them were inevitable but the difference in social status between the women and the men has persisted to the present day. That is why we give them food separately. We are their social superiors." Thought I: "Tradition indeed dies hard", and, with this thought, the word "mésalliance" passed through my mind.

A settlement of the dispute between the two families, for whom the date had been specially fixed, was a simple and straightforward matter. The case had its origin in the fickleness of a young woman. She was about nineteen years of age and had recently married, but shortly after her marriage she preferred another young man.

Everybody interested stood before me in my capacity as a magistrate. The fickle one was in the centre, with her sari well drawn up and nearly covering her face. Her lover, with his parents, was on her left. The deserted husband, with his father, mother and a brother, stood on her right. I knew perfectly well that if I ordered her to return to her husband she would be ill-treated and would, sooner or later, run away again. Trouble would go on recurring. The present position seemed thoroughly artificial since the marriage had been arranged by her parents while she was still a child.

No Solomon, I put the obviously superfluous question: "Whom do you prefer, this young man (pointing to her husband) or that young man (pointing to the other)? Confirm your answer by pointing your hand."

Hiding her face, with the exception of an eye, and shuffling her feet slightly, she replied without hesitation: "I prefer this one" (pointing to her lover).

Noting her decided preference the husband immediately said: "I don't object to her going to that man provided he pays me the full costs incurred on the marriage, amounting to two hundred and fifty rupees."

"I will," came the prompt reply from the favoured one.

The amount was deposited at once on my table. I counted it, then obtained a stamped receipt and made over the money to the husband, after obtaining the signatures of all concerned on a document setting forth the terms of the settlement. I specified an early date by which the young woman and her lover should marry, the date of the marriage to be reported to me by the headman of the village.

It was evident that everybody was satisfied with the sensible compromise which money alone had effected so promptly. Three days later the headman of the village came to me and reported that the lovers were married in accordance with Thåru ritual. The deserted husband was one large smile indicative of the pleasing prospect of another wife without additional expense. Strife and misery had given way to peace and happiness.

The following day I was ready to accompany Dyott on another picture-recording excursion. The chances of persuading the blue-bull and his cows to go in the direction of the camera faded completely. The animals had become too wild and they stampeded all over the country the moment they saw the elephants. Effort after effort failed to keep them within the crescent-shaped formation.

I therefore turned my attention to black-buck, and after much scouting found a herd. It consisted of two buck and five does. One of the former had turned black with age; the other, still young, and brown in colour, was probably his son.

Adopting the same strategy as in the case of the spotted deer, it took a comparatively short time to persuade the herd to travel slowly within a rough semi-circle of elephants towards Dyott. But as they approached the camera something happened to cause them to take fright. This, however, turned out miraculously in favour of photography. The older buck, alarmed, and to increase his visibility, began leaping high into the air, the others following suit.

In this promising, spectacular manner they passed in front of Dyott. This time the camera did not jam, and excellent action pictures were taken. I had the pleasure of seeing this herd again, not on terra firma but in a film entitled *Wild Life in India*, with myself in the background.

Dyott and I were once again on speaking terms, so we drove back to camp to the accompaniment of a friendly conversation. When we arrived and told Vernay and Faunthorpe of our success they were delighted. But there was more to be done. In spite of their daily endeavours they had not succeeded in obtaining a really good specimen of a sloth-bear. Those they had shot did not come up to museum standard.

Faunthorpe and I conferred. He accepted my suggestion that we should take six elephants and go to a particularly dense part of the forest some distance behind the Sálukápur rest-house. I knew there was every probability of finding bears in this locality because I had killed two or three here on various occasions.

Two of the six elephants had howdahs, one of which was occupied by Faunthorpe and the other by myself. The remaining four were saddled with pads; their function was to go through the grass and disturb the bears, while we went along the left and right flanks respectively. The pad elephants maintained a slow, easy pace, keeping about seven yards apart, uprooting tufts of grass and using them as fly-flaps, for it was hot and flies of all descriptions were a pest. If there were any bears they could not avoid being disturbed.

We moved forward methodically. My howdah was on a magnificent elephant named Lachmá. She was the staunchest of elephants in an expedition. Whether it was an angry tiger, or a charging bear, she stood as firm as a rock, permitting a good aim to be taken. We beat through several grassy patches and at last came upon one in which the pad elephants disturbed a bear. Out it shambled, complaining with a series of successive grunts at being disturbed during its normal slumbering hours.

Quickly taking in the situation, it decided on increasing speed and ran as fast as it could to the right and in front of Lachmá. Her máháwat, an experienced hunter, had brought her to a standstill by a whispered command. The bear was only five yards from Lachmá, at not too easy an angle to shoot down at from a howdah. I felt the chance was too good to miss. Following the bear's quick movements with the sight of my rifle, I fired and it rolled over. Rising in a second it charged Lachmá. Not budging an inch, she met the charge and, with her left foreleg, kicked the bear in an effortless manner, sending it sprawling several yards in front of her. With my second barrel I gave it no time to rise again.

Faunthorpe, thoroughly enjoying the incident, came along chuckling and anxious to see the size of the bear. He decided it was a worthy museum specimen: it measured half an inch over six feet from its nose to the tip of its tail. When standing up on its hind legs this bear must have been over seven feet in height, a formidable opponent to any man.

Vernay was delighted when he saw the bear on our return to camp. Perfectly mounted by John Jonas, this sloth-bear was duly dispatched to occupy a large glass apartment in the New York Natural History Museum.

CHAPTER VIII

CATTLE-KILLERS

TALL AND POWERFUL, the white bullocks leant their weight against the primitive plough guided by the hand of the villager. They were valuable beasts, lovely in the sight of their master, well trained and knowledgeable in his ways.

The field they were ploughing was a large one, about forty yards long and twenty-five wide. On its eastern side it was fringed by a forest that continued downwards for a considerable distance in a valley. From the nearest edge of the field the village site was about sixty yards.

When his little son arrived with his midday meal, the cultivator unyoked the bullocks, placed a generous quantity of freshly-cut green grass in front of them, retired to the shady side and, settling himself comfortably beneath a tree, began to eat. At the same time he watched his animals picking up mouthful after mouthful of the fodder as they stood near the other side of the field. Peace and serenity ruled over all.

Charging out of the forest and springing, as only a tiger in his prime can spring, the striped killer landed on the back of the bullock and astride it, attempting to deliver a felling blow at the animal's head.

But his landing was not as perfect as he had intended. The bullock, to save itself, had begun to dart forwards causing the tiger to sprawl awkwardly over its back, and to cling precariously to his now galloping mount by digging every one of his eighteen claws deep into its body on both sides of the spine. The stampeding bullocks raced towards the village huts, and the would-be killer considered it discreet to leap away and not be carried into the middle of the hamlet. His uncertain poise on the back of the violently jolting bullock had handicapped him. He had been too far sternwards on his mount to strike effectively.

Momentarily rendered speechless by the tragic happenings before his eyes, the cultivator recovered his voice and shouted as loudly as he could to scare away the tiger. Followed by his son he ran to the village, only to see the mauled bullock lying dead in the road. Shock had succeeded where the tiger's blow had failed.

Called to the scene by the village headman I saw the two white bullocks in the roadway, truly a magnificent pair of whose prowess and value I well knew. I examined on the body of the dead one the deep holes inflicted by the tiger's claws; the blood was still bubbling in those on the hind-quarters where they appeared to be deeper. I looked at the dead bullock's companion; it had already sat down. A villager placed a pail of water before it but it declined to drink. It was breathing heavily, staring, with its head up, at its dead companion three yards away. The owner and his sympathizers brought it titbits, but it refused all of them.

At this stage of the tragedy I decided to mount my bicycle and pedal back to my residence, five miles away, for my gun, with the idea of beating the forest for the cattle-killer. When I returned to the village the second bullock was no longer sitting; it was lying on its side, dying. The shock had been too great for it. The owner had suffered a tremendous loss, the more so since, to enable him to buy the bullocks, he had borrowed money at a high rate of interest from a bazaar banker.

There was still sufficient time for a beat before the sun set, and with the villagers' ready co-operation a good portion of the forest was covered, but the tiger had disappeared beyond the orbit of the beat. For the time being there was nothing more to be done except to await any news of the cattle-killer's movements which the headman could bring me. I had his assurance of adequate numbers of beaters when necessary.

Over three weeks passed before I heard of the tiger again. He had killed a cow at another village, about a mile away from Bonderwadi. His victim had been out grazing with the village herd. Receiving a message I went at once to the spot, glanced quickly at the carcass, studied the surroundings, and decided to have a platform made in a tree about twenty yards away, and to wait for the killer. He had eaten the entrails of the cow.

The platform constructed was far from comfortable. I did not mind its small size, but the hardness of the seat on which I would have to sit for four or five hours was an item justifying a mild objection. Thin, very short billets were placed corduroy-fashion and roped together; on top, as a finishing touch, a sparse mat made of date-palm leaves was laid. The headman noticed my hesitation in ascending the tree. I explained to him that a blanket would be greatly appreciated. He sent for one to the village, and when well folded and placed on the mat made a sufficiently soft seat for a fairly long vigil.

Settling down about half past two, I reckoned I had about four hours of daylight, strong enough to see the sights of my rifle. I then dismissed the helpful headman and his friends with the instruction to talk and laugh loudly as they went back to the village. Their conversation and forced mirth died slowly away, ending in a dead silence in the forest, the kind of silence that would tell the tiger he could now safely approach the carcass and resume his meal.

While I strained my ears and eyes for a sound or sign of the tiger, a jackal, followed by another, appeared on the scene. Their incomparable noses, capable of picking up a scent a hundred yards away with the help of the slightest of winds, led them to the carcass. But they would not go near it. They were all-too-conscious that the carcass was the property of a tiger, for they could smell the scent he had freshly left behind. They had good reason to suspect that he might spring out from hiding at any moment and deliver a fatal slap.

But jackals will ever be jackals. The inquisitive pair were a strong mixture of persistence and extreme nervousness. They would take a step or two towards the carcass, cause a dry leaf or decaying twig to crackle under their feet, and then retreat hastily. They were intensely jittery. The slightest sound affected them. Next, they would stand and look round, and then repeat the performance of approaching the carcass. Hungry, perhaps very hungry, the motto "safety first" prevailed. They circled widely round, and then trotted away as they thought it time to get out of danger's way and seek fresh woods and pastures new.

Their antics, while they lasted, gave me amusement and helped to while away the time and shorten my vigil. There was no doubt in my mind that they had, nevertheless, made a careful mental note of the carcass and, having done so, would return in due course to whatever was left after the owner had satisfied himself. There was always something for such scavengers.

By the time the jackals departed, the stillness and silence were complete, setting the scene for the climax. It seemed too early for the killer to appear. I blessed the headman for providing a blanket; otherwise, it would have been utterly impossible to have remained on the platform any longer.

Many were the jungle calls and sounds. Their succession was familiar. Each had its value, indicating as it did the exact time in the daily routine of each of the forest folk. As the hands of the clock rotate each day to tell the time, so these jungle calls and sounds, evening after evening, act as pointers to the coming scene.

The scent of the tiger around that particular spot was too strong to allow any of the deer tribe to wander close and none came in sight. I found it a handicap that the locality harboured no monkeys to give me a warning of the approach of the tiger. I must depend solely on my own eyes and ears. The sun was nearing the horizon. I was beginning to get anxious lest the failing light should spoil my vigil.

Into these moments of growing anxiety broke the unmistakable soft padded tread of the killer on the dry leaves. At first I thought he was going to show too much interest in my perch above him, and I steadied my gun, waiting for a sudden move, but the urge for his meal was too great, and he came on, straight beneath my tree. He walked to the carcass and sat at ease on his belly near it, pausing before enjoying a meal, possibly an exceptional one. In this position he showed me his head, his neck and the whole length of his back. I could not have wished for a clearer target, or one that gave me a better choice.

With no qualms about a sitting target I gently squeezed the trigger. My soft-nosed bullet penetrated deep between his shoulders and neck, his head falling instantly between his fore-paws. There was no need for a second shot.

After waiting for a few minutes, I whistled to summon the waiting villagers. They came on cautiously, only quickening their pace when I assured them all was well. We loaded the tiger on to a cart drawn by two bullocks, and began the journey to my residence. I followed on my cycle. Whenever the bullocks got the slightest scent of the killer, their driver had the utmost difficulty in controlling them; a whiff would make them dash forward and almost off the road. It was only the driver's tight reins through their noses that kept them under control.

This was my first cattle-killer, and when he was unloaded I measured him; the tape showed nine feet five inches from nose to tail tip. Not a tooth was missing, and he had already demonstrated on the white bullock that all his claws were intact. To all the cattle in the villages bordering the forest he had become a serious menace. On his list, cattle were number one; the deer, ample in the forest, had become number two.

Cattle-eaters were less of a reality in the town bazaar, and the owner of the two white bullocks had some difficulty in getting a second loan from the moneylender. It was not until I went and assured the usurer that the man's story was absolutely true that he decided to give another loan and, on my pleading, at a lower rate of interest.

My next experience of a cattle-killer concerned neither a cow, nor a bullock, nor a buffalo, but a baby camel. It was old enough to keep pace with its mother on the march, and occasionally to frolic round her, as she and the other camels walked along with measured tread from camp to camp with my touring baggage on their backs. Camels are not endearing animals but this little one lightened many moments for me.

I was in camp at Rámgarh, a pleasant spot on the northern slope of the Siwálik range of hills. A forest, only fifteen yards away, continued up to the ridge of the range which figured for many miles around as a well-defined watershed, the drainage on the northern aspect feeding the Ganges (now Gangá) river, while the waters flowing down the southern faces helped to swell the volume of the Jumná river (now Yámuná).

The camels, eight in number, were tethered for the night close to the rest-house which I was occupying. The baby camel was allowed to wander about at will, returning to its mother from time to time. During one of these wanderings a tiger killed it and ate a large part of its stomach.

I was outraged by the sudden loss of this young thing and determined to shoot the killer. I had a platform constructed in a nearby tree, although in this case I was unable to manage a blanket as well. Dining early, I ascended the sparsely-branched stem with the help of a ladder, and took up my position on the small platform. The carcass was about eighteen yards away. I continued sitting in the cold moonlight till two hours after midnight, hoping most fervently I would get a glimpse of the tiger when he returned. But he never came back. In truth it was just as well he did not, because when I attempted to handle my rifle, preparatory to descending from the platform, I found my hands and fingers so benumbed by the intensely bitter cold that I could not pull the trigger. Had the killer returned he would have been able to eat as much as he liked of the carcass without the least harm befalling him. I would have been a mere spectator in the brightest of bright moonlit nights.

Ignoring man-eaters which are always cattle-killers, my experience prompts me to divide cattle-killers into the two categories of the excusable and inexcusable.

In the first group are those tigers which have been compelled by circumstances to resort to killing cattle. Such a situation may be brought about by a severe epidemic amongst the deer in the forest concerned, causing them to die in large numbers. I have witnessed such an epidemic, the particular disease being rinderpest. Another circumstance is brought about by hunters of the kill-anything type, and poachers who spare nothing. It follows that if a tiger's hunger cannot be satisfied in the forest he is bound to go out of it and attack cattle near the villages, or grazing along the edge of the forest. Since the transfer of power in India, in the year 1947, the herbivorous fauna in certain forests, with which I was intimately acquainted for very many years right up to the year 1956, has been almost completely destroyed. The rapid depletion of game has been alarming. Unless the strictest measures are introduced forthwith complete extinction is inevitable. Such depletion means reducing to a dangerous extent the natural food of carnivora. Consequently, there are bound to be more and more excusable cattle-killers, if not man-eaters.

The category I have termed inexcusable includes tigers

which inhabit forests well stocked with deer but which deliberately choose to attack village cattle, and also cattle engaged in draught work, such as bullocks, and buffaloes dragging timber-laden carts.

At one of my camps, named Chandan Chowki, very extensive exploitation of the forests was in hand. Many scores of carts, drawn by bullocks, were busy transporting timber from the forests to the railway station.

One day, while inspecting the felling of trees and their conversion into building timber, I came across a very old log about ten or eleven feet in circumference, over which an exceptionally big bullock had been dragged into a nearby thicket. It was one of a valuable pair owned by a carter. There was no doubt whatsoever that the deed had been done by a tiger; the log showed the clearest signs of the drag over it. Round about was heavy grass except where the dead bullock had been pulled along the ground, causing the grass to be flattened. I could only make out the faintest footprints of a tiger, too faint to decide whether it was a male or female.

I stood pondering how any tiger could have dragged the huge animal over such a log, then called several cartmen to the log and inquired if they would give me details. In their usual nonchalant manner they replied: "A tiger killed the bullock and dragged it into that forest", pointing in a southerly direction. They added that several bullocks had been killed from time to time in the same locality.

I had already received a letter from Faunthorpe, who was then Commissioner of the Lucknow civil division, asking me to let him know if there was definite news of a tiger in any locality as he was anxious to spend a short time in the forests with me. After observing the fresh signs on the log and grass I was able to send him a telegram immediately.

He arrived in very good time. We beat the forest, using a line of elephants. Such operations are by no means always successful, but so well controlled was this beat that when a pair of tigers appeared almost simultaneously, we managed to shoot both of them. They were very fine specimens, in excellent condition, the tiger measuring a little over ten feet, and the tigress eight feet eight inches.

The problem which baffled me at the big log was now solved. The bullock had been killed and pulled over it by the combined strength of two tigers. Both were habitual cattle-killers, despite the fact that the forest held a very good stock of sambur, spotted-deer and also swamp deer. They were evidently finding the draught bullocks relatively easy prey, involving the minimum physical effort. In their condition cattle-killing was definitely inexcusable.

The Máhánt Saheb of Dehra Dun, the spiritual head of the Hindu community in that city, and the administrator of large estates, made large by property bequeathed to his temple from time to time, owned an enormous herd of milch cattle. All dry ones—and they numbered hundreds—were sent out, about twenty-five miles away, to a block of forests open to grazing, through which the Suswá river flowed in a limited channel, the actual boulder-strewn bed being very wide. The vast herd used to be accommodated on the edge of the dry watercourse under the shade of groups of trees. When a cow calved it was sent to the city.

The whole herd was constantly watched by a tiger which took up his abode on the opposite bank covered by scattered trees and patches of grass. Many of the cows used to calve in the dry river bed. Shortly after a cow calved the tiger stepped out and snatched the calf from its mother. Thus did he live on very tender veal, and continued to do so for a considerable time.

I happened to be in camp about three miles away. A great friend of mine, the owner of a large estate in the Dun, and the possessor of five excellent elephants, had tried unsuccessfully to shoot the tiger which had sought cover in a patch of grass and refused to be driven out of it. So my friend came to me about two o'clock one afternoon and insisted I should help him by temporarily abandoning the official papers on which I was busy.

The bonds of friendship prevailed. Accommodated on his best elephant, a tall tusker, I went with him to the forest. Here we spaced the five elephants suitably apart, and went slowly through the forest. When about half-way a barking-deer gave the alarm. Straightaway the tiger left the grass and bounded back between my elephant and the next. My máháwat turned the tusker and took up the pursuit at his fastest pace. About a hundred yards away I thought I saw the tiger half-hidden behind a log.

The elephant, well aware of the matter in hand, moved slowly towards him. I needed to be within thirty yards to give me the target I required, but too soon the cattle-killer charged. The tusker stood its ground and I had no choice but to shoot. Both my bullets found their mark, and to this day his skin and head adorn my verandah.

Sheer curiosity caused us to return to the patch of grass from which my friend had found it so difficult to dislodge him. We went into the patch and found half of a newly-born calf. Enjoying such a very tender morsel, is it any wonder he was so loath to budge?

The forest in which this tiger lived abounded in several large herds of spotted-deer. Stalking and trying to kill one of these deer meant far more physical effort than waiting on the edge of a river-bank and watching for calves to be born, thereafter to saunter out on to the dry bed of the river and snatch them from their mothers. He preferred easy living and, having made the choice, became an inexcusable cattle-killer. As such he paid the inevitable penalty.

CHAPTER IX

TWO MILK-LOVING SNAKES WITH DIFFERENT MANNERS

THE VERY THOUGHT of a snake, let alone the sight of one, makes people shudder, but if you are not one of the shuddering sort let me tell you, very briefly, about the snakes of India. There are nearly four hundred different species actually known and named. We may be quite sure that there are many more yet to be discovered, especially amongst those which live in or near water or in trees. I repeat, in trees, because on one occasion when I was inspecting felling operations in a high-quality forest the sawyers cut a sál (*Shorea robusta*) tree over a hundred feet in height to convert the stem into railway sleepers. Down came the giant to the ground with a thundering crash. From the dense foliage of its topmost branches, there emerged a pair of snakes, bright emerald green in colour, their heads sharply defined like the ace of spades, bodies whip-like and each about five feet in length.

They had lived at that incredible height, perfectly healthy and normally active. In my many years of observation I had never seen snakes like those. I stood amazed as they unconcernedly disentangled themselves from the maze of branches and leaves, and glided gracefully away on strange earthy surroundings to ascend, at the first opportunity, another giant of the forest and there to spend their lives again together in a new but thoroughly congenial abode in the highest branches.

Speaking non-technically, the hundreds of different recognizable kinds fall into two simple groups; poisonous and non-poisonous as far as man is concerned. There is a remarkable feature common to both groups. Both contain very big as well as very small snakes. By way of example, in the poisonous class there is the occasionally aggressive hamadryad, known as the king cobra, about twelve feet long and nine to twelve inches in girth; while the smallest is the very deadly saw-scaled viper, about eighteen inches long, and as thick as one's little finger. In the non-poisonous category the largest is undoubtedly the constricting python measuring up to twenty feet in length and as much as two feet in girth; in this group the smallest is the tiny, very shy grass-snake merely a foot in length and also as thick as your little finger.

Midway in size in the poisonous class is the ordinary, fascinating cobra. Its counterpart, in the non-poisonous group, is the dhâman, a great rat-killer, and therefore indirectly useful to man.

On the Indian plains I found the evenings in summer oppressively sultry. The atmosphere indoors became positively stifling, and compelled one to go for a drive or pay a visit to brother officials and sit with them in their gardens, discussing world events over a glass of cold beer, but inevitably returning to shop talk, with emphasis on the evergreen subject of the gross inadequacy of pay and pensions, so incommensurate, to our minds, with the responsibilities borne.

It was in such circumstances that I walked over uninvited to my friend the district Superintendent of Police. I knew I was always welcome for we had much in common in the study of Nature.

We strolled about his extensive compound, admiring the many species of birds and listening to their twitterings. About thirty yards to the left front of his bungalow there was a patch of scrub jungle. As we approached the edge of the mass of bushes I noticed a brown snake moving along under them.

"Did you see that snake?" I inquired of my host.

"Yes, I did," said he, casually adding: "that's my tame dhãman."

"Your tame dhãman?" I interrogated sceptically, being sure he was joking.

"I mean it," he retorted. "Come earlier tomorrow evening, punctually at six o'clock, and I'll show you how tame it is."

He had succeeded in rousing my curiosity which was just what he wanted to do. To my questions concerning the exact degree of tameness, and the methods employed to tame it, he was annoyingly unresponsive. I concluded he had a surprise in store for me, and seeing that he was enjoying keeping me in suspense, I asked no further questions. We returned to our chairs, placed well outside the bungalow to get away from the heat-radiating walls. We chatted for some time, and when I rose to leave he said smilingly: "Don't forget the dhãman, come punctually tomorrow evening and see."

The following evening I repeated the visit to my friend, taking particular care to be punctual. He was relaxing in a long easy chair and beckoned to me to sit in another. In preparation for the demonstration they were placed just as they were the previous evening. I had made up my mind not to say a word about the snake until he broached the subject. We next shared a bottle of beer. While we were talking about the day's doings he suddenly turned towards me and whispered: "Hush, keep absolutely still and silent, not a move. The dhãman is coming."

Our chairs were very close together. His was on the scrub side which meant that he could see what was happening. For the time being his body eclipsed everything in that direction. He repeated his injunction in the lowest of whispers: "Remain absolutely still." I obeyed him implicitly, even doing my best not to move an eyelash, but mounting curiosity made me look through the corners of my eyes to the left. Fortunately, I had had much practice in remaining still and silent, staring fixedly for long periods to the right and left observing the habits of animals and birds.

The dháman had many yards to travel from the bush. When it first emerged I could not see it behind my friend's substantial figure. But when it came nearer I saw it distinctly. On it came, propelling its long sinuous body in a series of graceful curves, producing an effect of superb rhythm. Not for the fraction of a second did it cease shooting out and retracting its forked tongue, so infinitely delicate in structure, so all-important as an organ for feeling its way, sensing its surroundings and functioning in perfect unison with its gleaming, bead-like eyes. Twice it halted, but only for a second or two while it glanced to the right and left, turning its head slightly. It then resumed its journey towards us.

As it came closer and still closer, I imagined every possible kind of situation, but there was no snake-charmer, only my host sitting in his chair, not a limb moving. I pictured him coaxing his tame dhaman on to his lap and then, entertainingly, passing it on to me.

The dháman was now right up to his chair. It then paused and, as I looked down at it, I noticed something I had not seen before. It was a saucer full of milk. The snake's head was now over the saucer and it drank till there was not a drop of milk left.

It then turned slowly away to begin its unhurried homeward journey. It was an effort to sustain my role of stillness and silence, but it was not for me to open the conversation. We watched the retreating and satisfied dháman. When it had disappeared amongst the bushes my friend turned towards me and triumphantly said: "It comes every evening, and at the same time, as if it has a watch."

"But how on earth did you train it for that performance?" I asked my smiling friend.

"I saw it one day in the scrub," he explained, "and later on I saw it again, so I concluded that this particular tract of bush was its home. I had heard from some Indian friends that snakes would drink milk and although I disbelieved them I thought I would like to test the idea with this dhámán. I started off with a saucer of milk placed just inside the edge of the scrub. When I looked at the saucer the next morning it was empty. But who had drunk the milk? Was it a jackal or a mongoose or my dhámán? I decided to repeat the experiment, but this time I put the saucer a yard or so outside the scrub in the direction of my bungalow. I then sat in the verandah and, with the help of my binoculars, kept watch on that saucer intently for nearly an hour. At last I saw the dhámán approach it. The sight intrigued me intensely and as soon as the snake moved away, I went quietly to the saucer and found it empty. My next plan was to try and attract the dhámán to me while I relaxed outside in the evenings. I needn't tell you how much perseverance this called for. Each day I placed the saucer of milk slightly further away from the edge of the scrub and nearer the bungalow. The dhámán, in its eagerness for the milk, disregarded the distance to be crawled and came closer and closer. In this manner and after many evenings I led it right up to my chair. I hope you have liked the demonstration in snake charming. You and I now know that, at least, a dhámán does drink milk and relishes it, and I must apologize to my Indian friends."

I congratulated my host on his admirable patience and success. Evening after evening he and I continued to observe the trusting brown dhámán come, enjoy its meal of milk, and return, leisurely and regaled, to its scrubby abode. For us, two lovers of Nature, the snake's daily visit was entertainment both rare and interesting.

When I received orders of transfer and was directed to take charge of the Central Almora forest division in the Himalayas, I rejoiced, as I was well acquainted with the territory and its likeable people. For the mountains, with their enchanting forests of pine and oak, I had a deep and enduring love. This unit was subsequently amalgamated with the Ranikhet division, the result being termed West Almora division which I administered for six years.

Almora, the capital of the district, was a quiet little station, eighty-one miles by motor road from railhead. It was 5,500 feet above sea level with an unsurpassable climate and an average annual rainfall of forty-two inches. The district as a whole had an infinite range of climates, varying from 2,500 feet above sea level and typical of the valleys, to the border line of perpetual snow where the high-level oak, spruce, silver fir, the white-flowered rhododendron, junipers and gentians cover the mountain sides. In these truly delightful conditions life was exceedingly pleasant, and well worth living after five years spent on the plains in the role of a much sun-tanned and heat-dried bureaucrat.

Arriving in Almora in the late autumn meant proceeding immediately on tour. Renting a bungalow was not, therefore, an urgent necessity. It was convenient to store my packages in one of the office godowns. But when July came, and touring ceased during the rainy months, a bungalow was essential and I accordingly rented one called Brighton House, presumably so named because it overlooked the deep-gorged Kosi river which, to the owner, was a substitute for the sea.

General stores, groceries, and mutton of fair quality were readily available. The supply of pure milk, however, presented quite a problem. The milk sellers, who came into the town daily from the surrounding villages, supplied milk of a notoriously bad quality, deliberately adulterated with water taken indiscriminately from any handy source. I tried several milkmen, one after another. Each swore on the most solemn of Hindu oaths that the milk would be absolutely pure on the following day. But the best that ever happened was a slight reduction in the proportion of unwanted water. One of the sellers of the adulterated product even told me that it was an inseparable part of a milkman's dharam (religion) never to supply pure milk. He continued, wholly unabashed: "So long as the resulting liquid has a milky appearance we consider it to be milk."

This candid confession set me a problem the only solution to which was the immediate purchase of a cow which had recently calved. The available animals were of very poor quality, but after looking at several, and feeling the utmost pity for all of them, I selected a white one which, even according to its owner, gave no more than a meagre four pints a day.

She had yet to attain middle-age and, like the majority of hill cows, was dwarfish. The purchase price was sixty-five rupees, equivalent approximately to five pounds. I named her Bessie, silently assuring her that I would give her good food and a kind home.

I appointed my gardener's son, Bir Singh, aged twenty-five, as Bessie's whole-time attendant at a monthly salary of fourteen rupees. A spacious stable in the outhouses was allotted and Bessie made comfortable with a generous quantity of bedding. Twice a day she was given a wooden-box full of a mixture consisting of two-thirds bran and one-third crushed gram, after soaking the latter in clean water for several hours. Arrangements were made for a sufficient quantity of fresh green grass to be brought daily from a village called Katihári, less than a mile away and situated on the northern aspect of Almora hill.

Under this generous treatment Bessie rapidly improved. Her ribs were no longer visible. With her hind-quarters and her hip bones now well covered with flesh, she assumed the shape of all well-cared-for cows. Automatically her lactation improved, eventually attaining quite a respectable figure for an animal of her miscellaneous breed, which not only satisfied my needs but also met her heifer calf's daily quota.

All went extremely well for about two months, until one

morning I had a report that she had failed to give the usual quantity.

To my searching inquiries Bir Singh's father returned an assurance that his son had not stolen the milk, adding: "We hillmen do not steal. If I and my son want a little milk we will ask you for it." In any case I knew that the orderly stood over him at milking-time. Inevitably my mind turned to the nearby villages, especially as the stable had no door and was easily accessible to anyone wishing to pay a nocturnal visit to Bessie, particularly her late owner, but when I openly voiced these suspicions there was a chorus of denial, emphatic that hillmen would never stoop to such dishonesty.

To me there seemed to be nothing wrong with the cow. She was eating as usual and I could notice no change in her condition. Thinking matters over I decided it was best not to take any risks so I dispatched a messenger to the local government-paid veterinary doctor, in case Bessie had early symptoms of some kind of bovine illness, and was very pleased at his ready response.

After a long and painstaking examination he said: "I cannot find anything wrong with her, and am unable to say why her lactation should drop. I'll make up a mixture and send it. In the meantime, all I can advise is to cut down her gram to half the present quantity. Particular attention should be given to washing the grass when it arrives from the village." The mixture arrived, and we duly administered it through a hollow bamboo, a method well known in Indian villages, but during the next few days there was no improvement in Bessie's condition.

The veterinary doctor came again, and this time he said he was not at all satisfied with her condition. Her appetite was not the same as before, and she seemed to be extremely nervous. I visited her in the morning before going to office, and then again as soon as I returned. Bir Singh and I would stand together looking at her, thinking hard, but quite baffled as to how we could bring her back to her normal condition.

While discussing her condition one evening, Bir Singh said: "Bessie is very fidgety at night. Since you changed my room and placed me next door to her stable I can hear her moving about a good deal." My orderly, a most observant man, who was standing nearby remarked: "I think somebody comes at night, milks the cow, and then quietly goes away. There is no door to the stable, so it is quite an easy matter to steal the milk. I say this because I can see the marks of ropes with which the thief tightly binds the cow's hind legs while he is milking her." Stooping down, he continued: "See, here are faint marks of the rope."

On hearing this I too stooped down and examined Bessie's hind legs. Right enough there were signs, though they were not very distinct. The orderly's remarks carried conviction. I turned to him and Bir Singh. "We must catch the thief," I said.

"The villages are quite close and it must be someone from there." The orderly agreed with my opinion, adding that the villagers sold milk in the town and were doing good business. This brief discussion opened up new avenues for thought. I ordered Bir Singh as soon as he heard Bessie fidgeting at night to come at once to my bedroom door, and wake me.

That same night Bir Singh gave a succession of knocks. I jumped out of bed, and warning him to follow me, took the hurricane lantern, and proceeded very stealthily to Bessie's stable, going right round, to avoid any chance of the thief seeing me first. The moon had disappeared behind a large black cloud. I lowered the flame of the lantern as much as I could. Then silently and swiftly moving along the side-wall I dashed to the front. I turned up the flame of the lantern, and with Bir Singh by my side with a stout stick in his hand, entered the stable all set to catch the thief with his pail and rope. But we found no thief, and I could only conclude that he had gone before we arrived. Bessie had stopped fidgeting, and seemed unperturbed.

I was determined to try again, and this time I changed my plan, ignoring the discomfort and loss of sleep I would suffer. I told Bir Singh I would occupy his room for the night and that I would stay awake with him. We would then hear Bessie fidgeting, and being next door with only a wall dividing the stable and ourselves, we would be in a much better position to come to terms with the thief. There would be no loss of time through Bir Singh coming to the bungalow and then returning to the stable with me.

I dined at the usual hour and then gave instructions for an easy chair to be taken to Bir Singh's room. I had the lantern replenished with oil, then walked across and settled down in the chair to await the sounds in the stable. On this occasion the moon was clear and bright.

I sat in the chair waiting. All I could hear was Bir Singh breathing. It was near eleven o'clock when Bessie began to fidget. I had taken care to keep Bir Singh's door wide open and to wear rubber-soled shoes. I listened intently. Bessie was indeed fidgeting once more. I approached the door as silently as I could. With the flame of the lantern at its highest I dashed to the stable door and looked straight in the direction of the cow's hind legs and udder, expecting to find the thief seated in the usual attitude adopted by villagers, with his thumbs and forefingers on two of the teats, milking into the pail held between his knees. Instead, an absolutely incredible sight met my eyes. There was a dhâman hurriedly uncoiling itself from Bessie's hind legs. In a flash it had disappeared.

Now I understood the cause of the rope-like marks on the cow's hind legs and told Bir Singh to call the orderly. When he came, I explained what had happened and instructed him to feel Bessie's udder. He said: "The dhâman has been milking her. Her teats are quite wet." The problem was undoubtedly solved. The dhâman was the milk thief and I had discovered it in the very act. I directed Bir Singh to untether Bessie, to tie her in his room and lock the door. By opening the window in the back wall there was enough air for the cow. I told the orderly to allow Bir Singh to sleep in his room for the rest of the night. After seeing that the cow had sufficient fodder, and the door was properly closed, I returned to the bungalow.

After breakfast the following day, before going to office, I decided to search for the dhâman. At about ten o'clock I called both the orderlies, Bir Singh and the groom, telling them to arm themselves with sticks. We would search around, find the dhâman and kill it. In one corner of the stable there was a large bundle of grass which the orderly suggested we should search first. As we proceeded to pull the bundle to pieces with our sticks Bir Singh shrieked: "There goes the dhâman." It had suddenly emerged from a hole in the left corner of the stable. We ran after it as it quickly made its way down two terraces of the garden where Bir Singh's father was weeding. We called to him to see where it went. Standing up immediately, he tracked it into a hole on the middle terrace. Instantly we gathered round the hole and were deciding what to do next, whether to dig it out or smoke it out, when the gardener suggested that it would be interesting to summon a snake-charmer whom he had seen the day before performing before a group of people in the bazaar. An orderly was dispatched at once to the musafir-khana (travellers' inn) as being the most likely place where the snake-charmer would be.

I was glad to see the orderly return, but he had with him a mere youth of nineteen or twenty years of age, with no equipment. Hearing why I had sent for him, he said he was an apprentice and that his master was having his breakfast. He then volunteered to return and tell his tutor to come and charm the snake out of the hole.

After about twenty minutes the master arrived with his full equipment consisting of two baskets laden with his performing snakes, and suspended from a long split resilient bamboo slung across his shoulders. Advancing towards me, garbed in saffron-coloured clothes with a turban of the same hue, he saluted. After returning his salutation and explaining the reason for summoning him, I guided him to the hole.

He studied it for a moment, then walked ten yards away and proceeded to unburden himself of the baskets which he deposited gently on the ground. Removing his turban he showed a well-cared-for head of hair, rather long but very neatly parted down the middle. He swept it back with his hands. He was in his late thirties, somewhat less than average height, slim, very wiry with a dark complexion made still darker by constant exposure to the sun. His deep-set black eyes had a quick, penetrating look about them. He next took off his shirt and tucked his dhoti (loin cloth) into his groin, pressing the folds firmly in. He was barefooted. All he wore now was the tucked-up loin cloth and a necklace made of tiny wooden beads hanging almost down to his rather narrow chest.

Picking up his gourd pipe, he sat down to face the hole with his toes only a yard from it. His posture was such that the entire weight of his body was on his feet, his buttocks being inches off the ground. This attitude enables a man to rise quickly. Playing the weird music, typical of all Indian snake-charmers, his cheeks bulged to their utmost extent, the veins on his neck and forehead swelling almost to bursting point as he increased the volume of sound.

He continued to play on for about half a minute with his extraordinarily keen eyes fixed on the hole, then as if on springs he suddenly shot straight up into the air. The snake had completely surprised him by emerging from an unnoticed hole only a foot away from him. Its emergence was a mere streak. Away it went along the terrace as fast as its five feet of length could wriggle, and its speed was truly incredible. Bounding along like a deer, the snake-charmer followed it and a race ensued, the dháman making a bee-line for a clump of bamboos at the end of the terrace. As he caught up with it, the snake-charmer, never slackening his pace, stooped, and with his right hand caught the reptile by the tail. There was no fumbling in spite of the feat being performed while both he and the dháman were moving at speed.

As soon as he gripped the tail he straightened his body, swung the snake round and round at arm's length above him, and then brought its head violently to the ground with a resounding thud. I marvelled at his dexterity. The unconscious dháman hung from his hand completely limp like a thick, brown rope. There was no attempt to wriggle, and the bifid tongue was completely withdrawn.

He walked slowly past me, as if the feat was an everyday occurrence. There was not the slightest sign of bravado or swank about his gait. Not a glance did he give me and, holding the senseless snake firmly by the tail with his right hand, and still carrying his pipe in his left, he went to his baskets. We had seen an admirable demonstration of how, with the aid of this weird music from a gourd pipe, a snake could be persuaded to quit its concealment and, with agility on the part of its pursuer, caught and stunned in a thoroughly effective manner.

We followed him to his baskets to see what he would do with the captive. He opened the lid of one and I saw coils and coils of reptiles. By their colour I recognized them as cobras. Pushing the dháman in, and forcibly coiling its body to make it fit into the basket, he closed the lid. The senseless snake was so limp that it was unable to coil itself.

He put on his shirt and, with both hands, again swept back his jet-black hair, after which he lowered his loin cloth. His turban lay an untidy heap on the ground. Unravelling it, he wound it round his head, and the exceptional care he took in doing so told me that a turban and the manner of its tying on the head had a special significance with him. When

On the left, Gulab Pyari, the one-time shaker. Note her inquisitiveness

The big tiger which mauled the elephant's trunk. The author (left) with Harry Young. Note the terriffic striking power in the forelegs

The author *(centre)* directing a tiger beat. The elephants are completely submerged by tall grass

A suspension bridge in the Himalayas, made by villagers out of grass twisted into ropes. The footpath is only ten inches wide

completely wound it gave his whole appearance quite a finishing touch, being about three times the size of his head. Its shade of saffron seemed to match his black, shining complexion.

Looking at me as much as to say "Now I am going", he lifted his baskets. "Not yet," I said. "Come up to the bungalow and give a performance with your cobras." I knew my men would be interested. He saluted and appeared only too glad to do so. Arriving near the porch I indicated a spot about two yards from the edge of the verandah. He lowered his burden, placing a basket on either side of him. I settled myself comfortably in a low chair, my men sat on the verandah floor near me. Bir Singh had already run as fast as he could to Bessie and her calf, and having replenished their green grass was back in good time to see the performance.

The snake-charmer, playing his pipe, brought out cobra after cobra from his baskets. There were three on his right and three on his left. All of them were swaying gently with their hoods expanded as if they were doing their best to keep time with the music. In succession he would move one and then another as a warning to maintain their erect positions and expand their hoods to their limit. Any cobra at all lethargically inclined was given a sharp flick of the hand to wake it up. When he returned a performer to a basket he would bring out another.

Continuing the music, and holding his pipe with his left hand, he stretched out his right to take out yet another cobra, but instantly withdrew it. Something unexpected had happened. "What's wrong?" I inquired. "The dháman has bitten me," he replied, at once clamping down the lid of the basket and losing no time in returning the performing cobras to the other basket. Indicating the last joint on the middle finger he showed the punctures caused by the dháman's teeth.

Rummaging quickly amongst his paraphernalia he found

a little bag, undid the knot at its neck and extracted a small object that looked like a stone but was shaped like a bean. He pressed it hard against the punctures and it stuck to them. Then he sat perfectly still and stared down at the ground. Very soon there were symptoms of distress. Sweat began to show more and more on his forehead. Again and again he drew the forefinger of his left hand across his brow and wiped it clean of sweat. Becoming anxious, I asked him if he would like me to send for a doctor. He negatived the idea with a slight shake of the head, so slight that I concluded that he was not fully conscious.

I could think of nothing but brandy. Hastening to my cupboard I poured out a stiff dose and brought it to him, telling him what it was. Hesitating at first, he gulped it down. The sweating continued and we observed him with anxiety. After about twenty minutes he appeared to be better and the sweating was less. Then with the forefinger and thumb of his left hand he removed the bean-like "stone" from the punctures and squeezed it. To our utter amazement, out fell drops of a liquid, light amber in colour—the poison the dháman had injected into him. Clearly the thing he applied had sucked the venom from his finger. I could hardly believe my eyes and all around me there were exclamations of wonder.

He was feeling much better but not quite well enough to pack and go away, and on my advice he stayed longer. I suspected he was also feeling the effect of the generous dose of brandy. So he remained sitting, and I continued observing him.

I have never known an adult human being to die from a dháman's bite, but it may prove fatal in the case of a child. I have seen instances when men and women, weeding in their fields during the rains, have been bitten by dhámans accidentally trodden upon; but with not a single fatal case amongst them. The bite certainly has the effect of making the victim feel ill for some time. This species is not aggressive.

Its habit is to escape as quickly as it can, and when doing so its speed is equal to that of any other Indian reptile.

I was naturally curious to learn what the black, bean-like "stone" was. Never before had I seen anything like it. To possess one would certainly have been an advantage in case of necessity. The snake-charmer regretted that he had none to spare. I did not press him, for I appreciated that, as a snake-catcher, his was the greater need. Replying to my next query as to where I could obtain such a stone, he said: "From the head of a large toad."

I next inquired what he would do with the now fully-awakened dhâman, because I was aware that this species is not usually kept by snake-charmers, whose favourite is the hood-expanding cobra. The dhâman has no hood to expand, and therefore does not give a showy performance. "Since it has bitten me," he said, "I must keep it and release it at Hardwar, our holy city on the Ganges river." He added: "Any snake that bites us, we snake charmers give to Mother Ganges." Hardwar was about three hundred miles away, but this distance was nothing to him since snake-charmers travel on foot from village to village, performing and earning. He went on: "I will reach Hardwar in about six weeks, in time for the big fair."

He was now ready to depart and I rewarded him handsomely. Bowing low, and touching his forehead with his right hand, he bade me good-bye. He had earned my highest admiration. After the day's doings I concluded that he ranked high among his colleagues. Doubtless, like many of them, he possessed this secret remedy that lay in a toad's head.

After his departure I went to Bessie's stable and proceeded to examine the hole out of which the dhâman had come. I was perplexed as to how the five-foot-long snake had accommodated itself without being seen, so I directed Bir Singh to bring a pickaxe and dig. As the digging continued the entrance widened, and then the mystery was solved. The hole led into a spacious chamber which had been made by rats to rear successive families. It was large enough for the snake to coil its body comfortably and remain unseen and snug. I had this chamber and hole filled with concrete and thoroughly rammed, and a close-fitting door fixed to the stable entrance.

It was now necessary to do everything possible to enable poor Bessie to recover from her dreadful nocturnal ordeals. Bir Singh, as anxious for her welfare as I was, insisted on sleeping in her stable for a week to restore her morale, and so speed her recovery. Given every possible care, and as much food as she could eat, she made noticeable progress. She put on weight and it was not long before she attained her maximum lactation.

Kind neighbours who possessed cows sent me my daily requirements of milk during Bessie's inability. Fortunately, her calf was old enough to graze and enjoy a soft diet. A few days later I related the whole remarkable incident to the veterinary doctor. "Well," he said, "I have heard tales of dhámans milking cows and even goats, but this is certainly the first case I have actually experienced." He was equally interested in the snake-charmer's cure. The "stone" in the toad's head has always intrigued me. There is an old volume entitled *Secret Wonders of Nature* by Fenton, written in the year 1569, in which the following interesting passage occurs:

> "There is found in the heads of toads a stone which they call borax or stelon: it is most commonly found in the head of a hee toad, of power to repulse poysons . . ."

Another author, named Lupton, who lived in the same period as Fenton, mentions the same belief in his book, *One Thousand Notable Things*.

That this belief in the efficacy of the "stone" was widespread as early as the sixteenth century there can be little doubt. It was actually shared by Shakespeare—who was a contemporary of Fenton—for in *As You Like It* Duke Senior says:

> Sweet are the uses of adversity,
> Which like the toad, ugly and venomous,
> Wears yet a precious jewel in his head.

CHAPTER X

LEOPARDS, BOLD AND CUNNING

VERY, VERY GENTLY the door was being opened. No one else in the room appeared to have noticed it; my hostess divided her attention between the tea-cups and her guests, while the four dogs sprawled somnolently at her feet. But even as I glanced at them the one nearest the door opened his eyes and I saw his hackles rise. His gaze was riveted on the slowly moving door. I must have missed the remark my hostess made to me, for as I turned my head to her she gave me a look of polite inquiry, which changed to astonishment as she, also, noticed the moving door. Into the room, within a foot of the now growling dog there stretched the silky paw of a leopard; groping inquisitively it curled round the door and beat gently on the floor.

In a single and comprehensive movement my hostess put down the best china, outflanked the dogs, and reached the door. In a voice full of indignation and authority she said "Shoo!" The leopard shooed.

Artistry requires that I now say we took up our tea-cups and went on with our interrupted conversations, but truth demands the confession that there was some considerable consternation and confusion as we took in the full impact of the intrusion. We all launched into anecdotes of situations of similar audacity and I was able to recount my favourite from my own experience.

This occurred when I was camping in tents about sixty yards from a village, and having retired for the night fell into a deep sleep after a day's hard trekking. My bed was against the wall of the tent, a position it always occupied. The wall was made of two folds of cloth, the outer being of stout material, while the inner, which served the purpose of a lining, was of a muslin-like texture, light yellow in colour.

Shortly after midnight I was roused by a persistent rubbing motion against the wall and along the length of the bed, against which there was a good deal of pressure. I imagined that a goat from the village had decided on rubbing itself, as goats habitually do, against the walls of huts or parapets of bridges. Not wanting the next move to be in the nature of a butting-battle, or even an attempt at devouring the tent fabric, I rose slowly and found my torch. I stepped outside and, after two paces, shone the light along the outer face of the tent wall. Into its thin beam stared the all-too-near eyes of a leopard. To our mutual surprise he instantly turned tail and was gone. Returning to the tent I pulled my bed away from the wall, and I remember giving the cloth a long and thoughtful inspection as I realized how little had separated me from the audacious creature rubbing his side to rid himself of irritation in the region of his ribs.

This characteristic audacity, combined with exceptional cunning, puts leopards in a class by themselves. They are definitely more cunning than tigers. Their colouring invariably blends perfectly with their environment, and when they lie on their bellies it is almost impossible to detect them.

As they stealthily stalk their victims with this low crouching habit and intensely eager expression, with green eyes glistening, they are the embodiment of deathly menace.

There are no characteristics justifying a division of *Felis pardus* into species, and they are therefore simply classified as the big and the small variety. Arbitrarily, sportsmen place under the big variety all specimens measuring seven feet six inches and over in length. The division into these varieties is based on numerous leopards shot and carefully measured; it is also consistent when habits and habitat are considered.

In height and length a big leopard is actually equal to a tigress. Specimens nine feet long have been shot, and very many between eight and nine. This particular variety is usually forest-loving and preys upon the same size of deer as a tiger does. If you see one sitting on its haunches in the middle of a forest road, as I have often done, and you are too far to notice the spots or rosettes, you are likely to mistake it for a tigress.

Except when a leopard of the big variety turns man-eater or cattle-killer, it generally prefers the depths of the forest to open lands sporadically covered with bush and scattered trees. As with all forest dwellers its movements are governed by the availability of food. Disease, followed by widespread mortality amongst deer in the forest, or increasing diminution of their numbers caused by that arch-destroyer, MAN, will compel a big leopard to quit his sylvan haunts and seek food outside them, but where the balance of Nature has been static I have known this variety of leopard to remain permanently in dense forest.

The small variety of leopard is better known on account of its boldness, and its habit of frequenting the immediate vicinity of houses and huts, prowling round them in the late evenings and at night with the object of seizing a calf, a goat or a stray dog. In the daytime it has no difficulty in finding ample cover in the adjoining scrub-jungle, ravines, rock-crowned hillocks and caves. In the Himalayas this variety kills ponies and full-size cattle, its clever strategy often taking the form of chasing and causing the animal to fall down a steep hillside, whence the leopard descends by a circuitous route, invariably to find its victim killed by the fall. It preys on flocks of sheep engaged in carrying borax, salt, and other commodities between India, Tibet and Central Asia; in the most daring manner it attacks these flocks at night in their barricaded pens at the various encampments by the roadside.

With the object of shooting a bird or two for the pot I shouldered my gun one evening and walked out of the Dholkhand rest-house, in the Saharanpur district. It was about half-past five as I went along a forest road, bordered on both sides by thickets giving ideal cover to all game.

After going about a mile I was intrigued to see a covey of wild fowl crossing the road and going in the direction of a wide fire-break. These birds did not appear to be unduly perturbed, though from the way in which the hens cackled at intervals I guessed that some animal was close by; I half expected to see a wild cat stalking them.

Instead, out prowled a leopard of the small variety, following the birds exactly like a cat. Crouching lower and lower he disappeared in the undergrowth. As he was lost to sight I ran forward to follow the sequence of events. The birds emerged on to the fire-break and evidently decided that it was no longer safe to continue on the bare ground since they were still being stalked. Without warning they rose in the air, but instantaneously the leopard sprang high, and striking out left and right with both fore-legs, succeeded in slapping a hen which he carried away.

Not many days later, in another locality, I watched a leopard pursuing some other birds in his most skilful manner. But on this occasion it was a clear case of cunning versus cunning. He resorted to every stratagem he could think of, but the birds moved on just ahead, keeping their eyes on him, and never attempting to fly right away. When they thought it time to fly up into a tree they did so, and sat on the branches looking contemptuously down at him. He, in turn, gazed up longingly at a peacock and three peahens. So unperturbed were these birds, and so conscious of their ability to out-manœuvre him, that they began preening themselves with great thoroughness and almost obvious enjoyment. They out-leoparded the leopard.

After completing my morning's inspections I was about half-way between my last camp and the next. Mounted on an elephant I emerged from the forest on to an external fire-break along which I continued for some way in order to satisfy myself that it was in an efficient condition. I had not gone far when I heard spotted-deer giving their calls of alarm in the adjoining scrub-covered lands.

Concluding that an animal was disturbing them I instructed the máháwat to drive the elephant very slowly in that direction and to halt when I advised him. We approached as silently as we could, taking cover whenever a tree permitted. There came in view an intensely interesting scene, one that I had never seen before during my many years in the forests, and I halted the elephant immediately.

A leopard was sitting on his haunches and around him was a ring of spotted-deer. There he sat in the centre, the distance between him and the deer being about fifteen yards. Their eyes were fixed on him, their hoofs were striking the ground, they were shaking their heads and flapping their ears. The leopard was a perfect picture as he turned his head slightly to the right and then to the left. Not a single deer took his eyes off him for a moment. He seemed to know that the slightest movement on his part would cause them to scatter in a way he did not want. On their part they knew that any relaxing in their alertness would prove fatal.

So he continued to study them hopefully, though pretending not to do so. He was waiting for an opportunity to seize one of their number. His little ears were being cocked in all directions as his peridot-like eyes continued to scan the circle round him. Losing patience, he at last changed his position and stood on all fours, whereupon the deer scattered in every direction. He made no attempt to chase them for in his clever brain there was a better idea. The picture of indifference, he strolled away.

Knowing his kind I continued to watch. He stood still and viewed his surroundings with particular reference to the scattered deer, now standing at a distance in groups of fours and fives. He chose the nearest group and decided to stalk them. The last I saw of him was when he was stealthily taking cover, from bush to bush, getting closer to the deer.

By stalking them his prospects were much better; sitting and looking at them, and they at him, with a safe distance between, he could never hope to catch one; at a getaway they would be much too quick for him.

How did such an interesting situation come about, a leopard with a ring of deer round him? I wish I had arrived a little earlier to see how it started. My guess is that the leopard and the deer, all residents of the same locality, were quite used to seeing one another, and on the strength of this acquaintanceship they knew that so long as they could see him they were safe. The leopard had shown himself in open lands. To prevent him stalking them they preferred, very wisely, to gather round and watch him, to put the candid question to him: "What are you going to do next?" Distance and speed were on their side only so long as they kept him under surveillance. Not to do so was to transfer the odds into his favour.

During one particular period of my career I developed an irrepressible urge for motor-cycling, to revel in the high exhilaration of travelling swiftly through the air without a windscreen in front or a hood above. The result was the purchase of a "Baby Triumph" two-stroke machine which developed the desired speed on the smooth-surfaced forest roads.

I continued to enjoy this longed-for exhilaration until one sultry afternoon which was anything but exhilarating. Speeding along to get a breath of air on a track with tall grass on either side, I turned a corner and came slap on two tiny leopard cubs playing in the middle of the road. It was impossible to stop before reaching them. In the same split-second I realized all-too-clearly that the fond parent was very near in the grass watching her offspring gambolling.

There was nothing to be done but to open the throttle lever wide, to give the little motor-cycle all the gas it would take and hope to avoid the cubs by flashing past them. As I came in line with them a black-spotted tawny-coloured body leapt out at me. Speed and speed alone saved me. One of the leopard's paws just missed my right thigh but struck the rear mudguard, causing the Triumph to develop a dangerous wobble. Recovering a life-saving equilibrium with difficulty I sped on my way, then drew to a somewhat shaky halt and glanced back. The leopardess was staring defiantly in my direction, her cubs motionless under her belly, and her tail switching vigorously, a sure indicator of her extreme anger.

For me the incident had a definite lesson: on such a small motor-cycle, excellent though it was in all other respects, there was no place for a gun. Sorrowfully, I sold it to a subordinate judge, and bought a bigger machine with a side-car capable of accommodating the weapons I needed to meet all circumstances. But the smaller motor-cycle gave me so much pleasure that I still remember its responsive throttle with feelings amounting almost to affection.

Leopards are arboreal. Regularly in the evenings I used to watch a pair, from the verandah of a rest-house, which overlooked a stretch of low-lying forest composed of miscellaneous species of trees, relatively short-boled with much-branched crowns.

About the same time every day both the leopards ascended one or other of these trees, which greatly increased their range of vision in the search for deer. When they saw a herd they descended with the agility of cats and began to stalk their prey. When successful they ate as much of the carcass as they wanted, and then, with admirable intelligence and foresight, carried whatever remained into the branches of a tree, thus depositing it securely out of reach of hyenas and jackals.

There is one animal which will cause a leopard to climb a tree as fast as he can manage it. That animal is the wild dog.

One day I was on an elephant, returning through the forest to my camp, and on reaching a point where four roads met, I took my usual careful look along them. On one of the side-roads I saw a pack of wild dogs sitting under a tree, gazing up into the branches. I wondered what it was they had trapped and were hoping to tear to pieces.

I made up my mind to kill as many of them as I could. Making a detour sufficiently wide not to disturb them, I reached a point about seventy yards away where I had the advantage of cover. Halting the elephant, descending to the ground, and directing the máháwat to take the elephant away, I approached the dogs as closely as I thought prudent, without disturbing them. They made a good target although I knew I should not get many of them. An absolute essential in shooting at wild dogs is a quick use of the bolt and magazine. Kneeling, I took aim and fired, killing two. Knowing the habits of wild dogs I waited for some time for the pack to return to devour the two dead ones. Stealthily they returned and I shot another one. The pack then disappeared.

Cautiously I went to the tree to investigate. High up in the branches was a leopard, as far from the ground as he could climb. This was his only refuge in such circumstances, for had the pack of wild dogs overtaken him on the ground they would have torn him to pieces and devoured him, though for a certainty two or three of their number would have perished in the attack. The survivors would have eaten these casualties.

A large pack of wild dogs will not, in fact, hesitate to ring a tiger. During the assault the latter will kill several of them but he is doomed to make a meal for the remainder. The presence of wild dogs in a block of forest causes all animals to scatter far and wide. Their method of killing has no equal in cruelty.

Of all man-eaters the leopard is the most audacious. It will even enter houses in its quest for human flesh. A Range Officer who served under me had a frightening experience when a man-eating leopard, which had already accounted for more than a hundred human lives, was roving and killing in his jurisdiction. He was at his headquarters at Jágeshwar, eighteen miles from Almora, a pretty place with a large grove of deodar trees amongst which nestled a highly sacred temple to which hundreds made frequent pilgrimages.

The Ranger, whose name was Nanda Ballabh Tewari, was a Bráhman of the highest caste. Wearing only a loin cloth and his sacred thread (janoee or yagyopavit) across his chest in the manner of all hill folk while cooking their food, he was sitting in front of the fire and throwing in more fuel when required. There was no chimney-stack to take away the smoke, so the door was left half open. It was his evening meal that he was preparing. Darkness was beginning to set in.

While thus sitting at ease he espied, in the dim light given by the fire and a hurricane lantern, the fearsome head of the man-eating leopard. Daringly, it took its first step into the Ranger's kitchen, to kill him and take his body away.

Never losing his presence of mind, the Ranger quickly picked up one of the burning billets and threw it forcefully at the murderous intruder. The flaming fragment found its mark, and the enemy retreated.

After the day's chores it was a form of relaxation to mount an elephant and roam in the forest in the hope of meeting a leopard or two and outwit them. Their extreme cunning made this attractive and entertaining. On one such occasion I had penetrated about three-quarters of a mile into the forest, progressing in a wide zigzag manner, making the elephant stand as still as possible for several minutes at a time. It is this practice of remaining motionless at intervals, and listening carefully, that bears fruit.

As the elephant stood, and I listened, I heard the frantic half-whistle half-yap of a spotted-deer to my right front, about fifty yards ahead. The forest was not dense but there were trees of medium size, and of many species, with low-spreading crowns beneath which there were shrubs of all kinds, making visibility very poor. There was, however, a single feature in my favour; a well-beaten track going in the direction from which the calls came.

I did not know whether I would meet a tiger or a leopard, and regretted I had not brought my double-barrel express rifle. As we drew nearer the elephant was made to walk at its slowest pace, and forbidden such noisy pastimes as uprooting tufts of grass or breaking twigs of trees.

The calls became louder. The elephant behaved perfectly and seemed to appreciate the situation. A few yards more and the scene opened up before me: a spotted-deer hind, with its little fawn hugging its side, was facing a leopard, which only ten yards away was creeping by imperceptible inches to get close enough to spring. Both leopard and hind were so engrossed watching each other that neither seemed to observe the soft approach of the elephant. The hind and fawn were in full view, but because of a tree with long-spreading branches, and tall bushes under them, I could see only the leopard's forelegs.

Every moment counted if I was to save the deer. My difficulty was to see more of the leopard, even a little more to get a worthwhile target. Legs, and only legs, were visible. If I was to obtain a glimpse of his chest it was essential, somehow, to lower the field of my vision and so look under the branches as much as possible.

There remained only one way of doing this: to lie as flat as I could along the pad. In spite of this new posture I still found my view was impeded. Life-saving seconds were passing, the leopard was closer, his spring closer still. I decided to aim with my trusty .315 B.S.A. magazine rifle at just above his legs, in the hope that the bullet would penetrate his chest.

The bullet hit him, but did not kill him. At the sound of

the shot the hind and fawn bounded away. The leopard, badly wounded, could not follow them, and it was now imperative to seek out and kill him. Telling the máháwat to drive the elephant forward to avoid the trees, I saw the leopard about thirty yards away, making his way to a group of bushes. He was now a good target and before he reached them I fired and dropped him. The first shot was nothing more than a time-dictated guess; but it had served a purpose.

The hind, weaponless, had faced a foe armed with fangs and claws advancing with deadly intent. I had witnessed a most vivid portrayal of the mother-instinct, as she stood, in mortal peril, steadfast to guard the life of her fawn.

A delightful memory—the incomparable majesty of the Higher Himalayas. In the centre, the three-peaked Trisul (trinity) 23,406 feet, each peak bearing the name of a Hindu goddess. On the left Nanda Ghunti, 20,700 feet. On the extreme right, Nanda Devi, 25,645 feet

The author with the man-eater of Gwaldom

he Bhotias' encampment raided by the other leopard. The tiny bags contain borax from Tibet and are carried of the backs of sheep

CHAPTER XI

THE MAN-EATING TIGRESS OF GWALDOM

SHE WAS AN old tigress quite incapable of catching the very nimble barking-deer in a mountainous tract of forest. Sheer hunger had driven her to prey upon men and women, and despite her age she had already killed more than twenty.

Most of her victims were women, as their daily routine involved them in the greatest risks. After the morning's chores, which included cooking the next meal, attending to the family's cattle, sweeping the house, and removing the dung from the cattle-shed to store it outside for manuring the fields, they picked up their sickles about nine o'clock and went in small groups to the nearest forest, there to cut grass and lop oak leaves; these they made into large headloads which they brought back to the village as night fodder for the cattle. After all this they would cook the evening meal. They were certainly the workers in the family.

During the ploughing season the men would be busy preparing their fields for the next crop, but when the crops were standing and gradually ripening, the husbands would congregate in the wayside shops to gossip, and meet in one another's houses to gamble. Very rarely would they accompany their wives to the forest. Thus the distribution of labour in these families, residing in the villages scattered over the enormous tract of the lower Himalayas and comprising the districts of Kumaon and Garhwal, was sharply defined and very much in favour of the men.

On one of my many tours in this vast territory I happened to be staying with a friend, Robert Bellairs, at his headquarters, Kausani, in the Almora district. On his return from the war he had immediately been appointed Settlement Officer to apportion out land amongst landless ex-soldiers and their families. There could not have been a better selection for such work since he knew the country intimately, as well as the men with whom he was dealing. He had raised a complete Labour Corps in this very tract and commanded it in Europe and the Middle East throughout the war.

He occupied a spacious bungalow, once the property of two brothers, Norman and Colin Troup, the former a Tea Planter, the latter a retired Captain from the Royal Navy. Both brothers died between 1916 and 1920. Their extensive estates, comprising tea gardens and orchards, were then acquired by the Government to be handed over, in the form of small but adequate holdings, to the hundreds of soldiers whose services were no longer required.

Though there were two travellers' bungalows at Kausani, Robert Bellairs would never permit me to occupy either. Immediately he heard I had arrived on tour he would promptly walk over to the bungalows and insist on my leaving, to stay much more comfortably with him. So over I would go as his guest, my camp staff being accommodated in some of the several vacant buildings.

It was in this manner that I heard of the man-eating tigress of Gwaldom. Robert and I were at lunch. He had finished hearing petitions for land from a score or so of ex-army men. I had returned from my inspections of adjoining government forests where pine trees were being tapped for crude resin, a rapidly expanding industry in those days.

We had scarcely risen from the table when two breathless runners from Gwaldom were announced. Stepping out on to the verandah we listened to their urgent report. They said another woman had been killed. She had been taken by the killer when she and her companions from a village near

Gwaldom had gone to the jungle to cut fodder. While they were returning, well before sunset, each with her head-load, the man-eater had pounced upon one of them. Gripping and killing her, the tigress rolled the body downhill, then dragged it into the undergrowth to devour.

To our many inquiries they described the place where the tragedy had occurred. Both Robert and I were well acquainted with the forests adjoining Gwaldom; we had hunted and shot leopards in that locality. Our method was a strange one: we would sit on the ground back to back. We found this unique position quite interesting and one that met with much success for it increased our joint vision to a complete circle; sometimes a leopard would appear in front of him, at other times I would have the luck; never did we allow one to pass alive.

After hearing the account of the tragedy and concluding that the man-eater would still be in the Gwaldom forest, we decided to move camp the following day, and instructed the two messengers to warn the villagers not to go to the forest. Robert had work in the vicinity and I would have an opportunity of seeing felling of pine trees by one of my contractors. We could thus profitably combine our work with an endeavour to destroy the menace.

Robert, an excellent rifle-shot, was anxious to make an early attempt to contact the man-eater particularly as it had already evaded him once before. He had sat in a pit near the remains of a cow which it had killed, erecting round himself a hedge of branches and leaves as camouflage. The man-eater, approaching very warily, saw this strange feature in the forest from a considerable distance and evidently became deeply suspicious. It circled widely round and round, taking cover cleverly in the undergrowth and then disappearing. Robert could hear its growls of disapproval but it never showed itself sufficiently to give him a chance of a shot. He climbed out of his pit, stiff and disappointed but determined to have another try.

There was no difficulty in moving camp the next morning even at such short notice. I had an excellent team of fifteen mules, powerful and well-cared-for, capable of transporting the whole of my camp equipment and office many miles in a day. These mules were my especial care for on their fitness depended my mobility and timely supervision of a variety of operations over a very large hilly area.

Looking to the north-west from Kausani, Gwaldom was plainly visible. There were no intervening hills, only the valley of Baijnath. From Kausani the bridle-path zigzagged fairly steeply down through pine forests to Garur, thence it had a gentle gradient to the lowest point where there was a large village with extensively terraced fields. From here the bridle-path ascended severely through open pine forests right up to Gwaldom.

At intervals the mules, with their full loads, would stop for a few minutes to rest, then the leader without the least persuasion would move forward and the remainder would follow with their grooms alongside. Never did any of these animals require the least touch of a whip to urge them onwards on any march.

The distance from Kausani to Gwaldom is about ten miles. When, through the crowns of the pines, our destination came in sight we saw a group of villagers waiting by the roadside. They were from the village near Gwaldom, to the north-east of it. These people were closely concerned with the tragedy; the dead woman was from their village and her husband was with them.

After the usual greetings the men gave further details of the occurrence and pointed to the forest where the woman was killed. We explained to them that we had specially come to try and destroy the man-eater, adding that if it was necessary we would ask them to beat the jungle; to this they readily agreed. It was very much in their interests that the tigress should be killed.

As they accompanied us to Gwaldom, they went on to

explain that all the women of their village were now terrified and refused to go again to the forest to bring back grass and leaf fodder for the cattle which, in consequence, were on very short rations. To this we replied that in our opinion the women were quite right, and that it was high time the men took their turn at going to the jungle for fodder. It did not surprise us however that they strongly disagreed; we knew that it was in accordance with their traditions and customs for husbands to look upon their wives as mere servants. There was a common saying amongst them: "The more wives, the more servants."

Gwaldom was truly a very lovely camping site. Situated on a prominent spur forming a watershed between two substantial rivers, the Pindar flowing to the west, the Gumti to the east, it was exactly on the boundary dividing the districts of Almora and Garhwal.

That famous mountaineer, F. S. Smythe, one of the conquerors of the 25,447-feet Kamet, has included in his charming book *The Valley of Flowers* a delightful colour photograph of the great foothills from Gwaldom in the afternoon light, though unfortunately the clouds which usually gather as the day advances have obscured the mighty snowy peaks.

Determined to meet the man-eater, we picked up our rifles and went immediately to the portion of the forest the villagers had indicated. We followed a very narrow foot-track that led through thickets of brambles and berberis. The villagers followed us silently but as we advanced further we thought it prudent to dispense with their company. If it proved necessary to stalk the man-eater we preferred to do it alone. We told them to wait, that we would give them further instructions and would ask for their help if circumstances made it essential. Amiably acquiescent, they squatted where we left them.

We entered a forest of oak trees every one of which was gnarled and misshapen, the result of continual lopping for leaf-fodder. A little further ahead we came upon a point where two footpaths met, one going roughly along the contour, the other leading southwards and up a high ridge. The entire jungle faced north, overlooking Gwaldom and the lovely vale of the Pindar river.

Stopping to reconnoitre we decided to take the narrow path along the contour and to follow it throughout its course. This was the direction the men had pointed out. The greatest vigilance was now called for to avoid being surprised by the killer. The jungle with its undergrowth made ideal conditions for a man-eater's abode. The tigress had killed only a fortnight ago and this woman was the second victim within that time; encouragement enough for it to continue in this particular spot.

As we progressed slowly we saw the first of the reliable signs we were looking for . . . bundles of grass and leaf-fodder thrown here and there by the women when one of them had cried the alarm "Sher!" (tiger). On we went for another sixty yards when we came upon the unmistakable scene of the tragedy.

The bundles were scattered about, two of them a little above the path, but most of them below, having rolled till they came to rest against bushes. Sickles were still securely fastened under the ropes holding the fodder together.

Viewing our surroundings very carefully we considered that downhill was far more interesting than uphill. So down we went, step by step, and side by side. All was so quiet that we sensed tragedy and the presence of the man-eater in the very atmosphere. We were now on the line of the drag through the undergrowth.

Stopping repeatedly to listen for the slightest sound and to observe any movement, with our rifles ready, we went forward as silently as we could, expecting to come upon the remains of the unfortunate woman, or fragments of her garments amongst the bushes.

We had not gone ten yards further when there was a violent agitation of the bushes to our right front, about twenty yards from us. Then stillness, and we heard low growls. It was the tigress. She had heard us and could possibly see us, but from where we stood we could not see her. We were acutely aware of this disadvantage.

It was essential to change our position and manœuvre slightly to try and catch a glimpse of her. Stepping very slowly a yard to our right so as not to lose advantage of our uphill position, as we expected her to rush at us, we still failed to see any portion of her tawny body. The growls continued, and grew louder. We then moved another step to the right when we both got a glimpse of her.

It was a chance not to be missed. Our rifles rang out simultaneously. The tigress fell, but recovered herself, and we lost sight of her as she disappeared amongst the bushes. We quite expected her to come straight at us, but we saw by the stirring tips of the bushes that she had gone downhill.

We now knew that at least one of the bullets had wounded her, and that the wound would tell increasingly as time went on. We were on the northern face of the high ridge to the south, over which the sun was about to dip. Looking at my watch I found it was nearly five o'clock. It was unwise to continue the hunt in an unfavourable light in which all the odds would be in the favour of the now infuriated man-eater. The idea of a beat so late in the afternoon was entirely out of the question; it meant risking lives unnecessarily. So we reluctantly called it a day.

Two objects had been definitely achieved; we had found the tigress and we had, we hoped, hurt her so severely that she would be unable to move out of the patch of jungle. The situation was full of promise for a beat on the morrow when we would be able to come to grips with her again finally.

Retracing our steps and seeing no sign of the dead

woman's remains or her clothes, we reached the path and stood again for a few minutes on the very spot where the tigress had attacked the woman. After studying the surroundings and getting an idea how best to conduct the morrow's operations, we walked back along the path and found the villagers waiting for us just where we had left them.

There were many inquiring and anxious looks. Telling them what had happened, we said there was every hope of killing the tigress during the beat if they would bring at least seventy beaters in the morning. They asked about the victim's remains, to which we replied that we would carry out a thorough search after the beat.

That night, after dinner, Robert and I picked up pencil and paper and worked out, by means of a sketch, just what might happen and the best way of preventing the tigress from escaping if she was still able to travel. If she was, then which was her most probable direction?

Towards the west were fairly extensive fields under cultivation. There were no trees and no scrub, which meant a complete absence of cover. We took into consideration the probability that the tigress knew every yard of the ground. In the north, and leading down to the Pindar river, the forest opened up considerably, affording very little cover; we concluded that she was not likely to choose this direction which would mean swimming the wide Pindar in full view of the pursuers.

So the problem really concerned the south and the east where both forest and undergrowth continued. On these two points of the compass our minds were accordingly concentrated. At the same time, as hunters experienced in the ways of wounded animals we knew that the south and the west could not be entirely neglected.

After discussion we retired for the night, hoping the villagers would not fail us, and that all would be well on the morrow with the destruction of the tigress.

Next morning the villagers arrived in force at seven o'clock. They had been able to muster sixty-two men all told. As we were about to start out, three retired Garhwali officers from the 18th Royal Garwhal Rifles greeted us, saying they had heard about yesterday's doings and had come to assist us in the beat. They carried good modern sporting rifles. These splendid men had put in their full term of service, had been through the war in France, and all three had been decorated for valour. They had settled down in their village homes about three miles from Gwaldom. They were introduced to me by Robert as Subedar Major Chaman Singh, Subedar Bhowan Singh and Jemadar Baldeo Singh. Robert knew them well. They were on his committee for considering claims and allotting land amongst ex-soldiers of their regiment.

The presence of five guns, instead of two, helped greatly to simplify the main problem of the east and south where the tigress was most likely to break out. We therefore stationed three guns on the east; one gun on the south, the uphill side, and another in the north in case the disabled tigress would only move in a downhill direction. Along the west we placed several men on trees with instructions to clap their hands loudly if they saw the tigress coming towards them and so make her change direction towards the guns.

We decided that the beat should be from west to east where most of the guns would be waiting. Baldeo Singh was on the north, Bhowan Singh on the south, Chaman Singh, Robert and myself in the east. To prevent accidents due to cross-firing we ascended low branching oak trees which, incidentally, increased our field of vision. We instructed the beaters to climb trees at very short intervals and look carefully ahead before moving forward.

With a great volume of shouting the beat began. It had not progressed half an hour when one of the beaters, from his perch in a tree, yelled loudly that he could see the tigress. Subedar Bhowan Singh who was nearest to him promptly descended from his tree and climbed up beside the beater. He called out that he also could see the tigress and that she was lying on her side.

He then called out to us again, saying he would direct us to the spot where the tigress was. Orders were given for the beat to stop and Robert, Chaman Singh and I got down from our perches, and went through the undergrowth; Bhowan Singh from his perch gave his directions as if he were on a parade ground, and we maintained a spacing of about five yards.

At intervals Bhowan Singh gave us approximate distances in addition to direction, adding that he would fire at the tigress if she moved. Then we saw her, lying quite still in a pool of blood near her head. There was little doubt in our minds that she was dead, but we approached cautiously to examine her.

Both the bullets, fired by Robert and myself, the previous afternoon had found their mark, one in the stomach, the other in the lungs; this latter shot had caused her to vomit much blood. In fact she had gone less than fifteen yards after being hit.

She was decidedly aged and in extremely poor condition. Her left hind leg appeared to be deformed and must have been a definite disability. Several teeth, including two canines, were missing. We measured her at the camp and found her length from the nose to the tip of the tail to be eight feet five and a quarter inches.

Before leaving the jungle we scoured the undergrowth and found the woman's scanty remains. These were collected by the villagers and taken down to the left bank of the Pindar river and there cremated.

Returning to Gwaldom we enjoyed an hour's conversation with the three Garhwalo officers, and on parting thanked them warmly for their assistance so ably given and so very welcome.

Our trip to Gwaldom had not been in vain but it was the

last of the many pleasant and thrilling adventures Robert and I had together. With his settlers now established, he decided to change the Himalayan skyline for the rare snows of Africa, and go coffee planting near the lower slopes of Kilimanjaro.

CHAPTER XII

THE MAN-EATING LEOPARD OF CHOWKOOREE

I FIRST HEARD of him in a manner which I thought truly unique, and, incongruously, quite amusing.

The time had come for elections to membership of the Legislative Assembly of the United Provinces (now Uttar Pradesh). Various centres were selected by the Civil Authorities of the Almora district as polling-booths. One such was Bageshwar on the Sarju, a river with its source in the high Himalayas. It was a sacred place, not only for the periodical washing away of sins during important Hindu festivals, but also for the cremation of corpses which were brought from all villages within a ten-mile radius.

Bageshwar was also a trade centre, where a large annual fair was held about mid-January. Thousands of men, women and children flocked to it and much money and merchandise changed hands. It boasted a permanent bazaar consisting of two rows of small shops with a cobbled street between them. These tiny establishments dealt in prodigious quantities of grain, richly-coloured cloth, and a haphazard assortment of hand-made agricultural implements. A lighter note was struck by the stocks of cheap jewellery brightened by red, blue, green and yellow beads, hung prominently in the shop front, and irresistibly alluring to the simple villagers.

For the purpose of the elections, and as one of the senior officials in the district, I was appointed Presiding Officer at Bageshwar. Under me were several vote-recording officials and the spacious rest-house was transformed into a polling-booth. In each of the rooms sat a recording officer to take the votes from voters who entered as their names were announced aloud from the electoral rolls.

A vote was a piece of stiffish paper bearing the printed names of the candidates. When a voter approached the recording officer's table the latter read out the names of the candidates and then put the question: "For whom do you wish to vote?" If literate, the voter put an X against his candidate, if not, the recording officer made the mark for him. The paper was then pushed through a narrow slit in the sealed lid of the ballot-box.

In a procedure of this kind much depends on the recording officer. The greater the illiteracy the greater is the need for absolute impartiality. The candidates' agents, who were permitted to move freely about the rooms, were deeply suspicious, with the inevitable result that there were frequent altercations between them and the recording officers. There were accusations that the latter were entering the X against the wrong candidate's name and then quickly slipping the vote paper into the ballot-box. There were only two candidates up for election, therefore only two agents, whilst there were several recording officers; this meant that the agents could not be everywhere; the candidates themselves were outside the enclosure doing their utmost to influence voters.

I walked actively from room to room supervising as best I could in these difficult circumstances, and continually pouring oil on very troubled waters. Like the agents I was severely handicapped, since it was physically impossible for me to be looking over the shoulders of all the recording officers at one and the same time.

While I was adjudicating on one of the numerous complaints lodged by an indignant agent and hotly refuted by the recording officer, an aged voter was ushered into the room to cast his vote. It was clear that he came from a distant village; bent and very dependent on his walking-stick he shuffled along to the recording officer's table. As he stood there he looked up at the three of us: myself, the recording officer and the agent. His critical old eyes were appraising each of us in turn, and he was paying scant attention to the recording officer's question.

Patiently, politely, and showing commendable respect for senility, Salig Ram Thapliyal, the recording officer, once more put the question in a loud voice to ageing ears:

"For whom do you wish to vote? There are two candidates and their names are: Anand Singh and Badri Datt Pande."

"I don't know either of them," came the reply, "nor do I know anything about them. I was instructed by the headman of my village, Chowkooree, to be present here today. I don't even know why there is a big crowd today in Bageshwar."

Still with great patience the recording officer explained to him the object of the meeting, adding that it was imperative he should choose one of the two candidates. But the aged one, a staunch believer in independent thinking, had his own ideas about elections to the Legislative Assembly.

"I don't want to give my vote to strangers," said he, "I will give it to this brave gentleman (raising his lean arm and pointing at me) who is standing here, because I know that he will come and shoot the leopard which has killed my nephew on my very doorstep, and brought great danger and unhappiness to me in my declining days."

Very fortunately I was in a position to answer this flattery by assuring him that my planned itinerary would take me in the direction of Chowkooree where I had tree-marking operations to inspect. It rather depended how far his village was from the scene of my work, but I certainly would do my best to shoot the man-eater so that he could indeed spend his days in peace and security. Hearing this his wrinkled face and bedimmed grey eyes immediately lit up and his old head nodded again and again with satisfaction. Leaning more heavily on his stick, he looked at the long-suffering recording officer and obviously having forgotten all about the candidates inquired:

"What more am I supposed to do here?"

I thought it best for all concerned to take on myself the task of explaining to him just what was required of him by way of civic duties. During my elucidation I noticed his eyes took on a very far-away expression, as if still picturing the prowling leopard in his distant village; reluctantly recalled by the sibilant tones of the recording officer once more pronouncing the names of the two candidates, he looked at us with a sudden clear smile.

"I would never vote for the Bráhman," he said, "I will give my vote to the Rajput since I am a Rajput myself; my name is Narbat Singh."

Accordingly an X was marked against the name of the candidate Anand Singh, the Rajput. As he walked out of the room Narbat Singh looked back at me and said, "I pray you not to forget, but keep your promise and kill the leopard." I gave him an assurance in the form of a big and consoling wave of the hand. Hesitating on the verandah he began an earnest conversation with the orderly on duty. This had to be stopped, and he was led gently by the arm out of the enclosure. As I watched his uncertain steps turn towards his village, I was overwhelmed by a sudden sense of obligation to this lonely old man, and I pledged myself to end the life of that man-eater, so that in truth he might feel secure.

When to everyone's relief the elections were over, the ballot-boxes, securely sealed, were dispatched to headquarters under police escort, and I was free to leave Bageshwar. I travelled by stages in a general easterly direction in accordance with the itinerary already notified to officials concerned with my movements.

I had never been to Chowkooree village but had visited a tea plantation by the same name, having on a previous tour called on its owner, Stevenson, who lived there with his family. Particularly did I remember a tea tree growing near the corner of his house. He had allowed a tea bush to grow into a medium-sized tree, which, of course, is its natural habit if it is not annually pruned down. Stevenson was well past his three-score years and ten, quite beyond hunting a man-eater. As the plantation bore the same name as the village and was in the same direction I concluded that it was within the village territorial limits in the official records.

On the fourth day I was passing along the road skirting Stevenson's estate and in response to a standing invitation, called in to see the family. I inquired from the old gentleman about the man-eating. "I have heard about it through my men," he replied, "but it hasn't killed anybody or done any damage on my estate. They say it is killing people round about Chowkooree village which is a good distance from here." I asked him how far, but he was very vague, and hazarded a guess of "about three and a half miles".

Bidding the Stevenson family good-bye I continued my march for the day and arrived at Berenag. Halting here for two days to attend to boundary disputes, I next went due north to the Thal rest house. Thal was about four hundred feet above the turbulent Ramganga river which was spanned by a modern and very substantial suspension bridge. I had descended from 5,900 feet to 3,300 feet above sea level. The rocky bed of the Ramganga had retarded timber-floating work so much that I was compelled to stay three days trying to encourage the floating-gangs to make better progress, but this was a tricky job as I could appreciate only too well the immense difficulties they were encountering.

My next march was in a north-westerly direction to Kalinag, which means the Black Mountain, about eight miles from Thal. Here the absence of a rest-house necessitated pitching tents. I settled down for a four days' halt to complete inspections of timber operations, the construction of a new road and, as I have said, the selection and marking of trees for the following season's programme.

Kalinag, though prettily situated amongst pines, was a bleak camp, and doubly so in the grip of winter. It overlooked the Ramganga and also the trade route that went north-easterly, winding through the forests on the opposite hills and eventually away into Tibet.

This tract of country was, in fact, the land of leopards, which found adequate concealment in narrow, bush-covered ravines and prowled forth at intervals to kill their prey, whether sheep or goat or, occasionally, a human being. Hill-born leopards are remarkably lithesome and agile on mountains. Like large-sized cats, tawny-coloured and black-spotted, they can make astonishing leaps from rock to rock, and always in the most graceful manner.

Here at last I had encouraging replies to my inquiries about the man-eater. Chowkooree village was now, at last, little less than Stevenson's "three miles" distant on the same side of the river.

I spent the forenoon of the first day seeing operations. While at lunch I decided that it might be profitable to devote the afternoon to a reconnaissance between my camp and Chowkooree village, which was westwards. Slinging a rifle across my shoulder I went along at an easy pace observing the general conformation of the terrain, the ravines that scarred the hillsides, and the extent and character of the cover in them. Many of the large rocks scattered about looked as though at any moment they would lose their precarious hold and come crashing down into the river.

After about an hour's strolling I came to the outskirts of Narbat Singh's village. The houses were close together on a flat piece of land and the terraced fields above and below were covered with their winter crops. Approaching the nearest house I saw children playing in front of it, deeply engrossed in throwing handfuls of mud at one another and then running and hiding behind the corners of the house to escape retaliation. The moment they set eyes on me they seemed petrified. They stood still, gasped, and then scampered away into the houses. I suppose they had never seen a white man before.

At the other end of the row of houses there was a group of medlars in whose shade sat four old men, smoking and passing a common pipe around. A sleeping dog rose, barked furiously, then retreated to the rear of a house to continue barking. A goat with a kid was tethered at the edge of the front courtyard, nibbling greedily at a bundle of fresh leaves placed before it.

The men, surprised to see me, touched their foreheads, each greeting me in turn and looking inquisitively. To allay their suspicions I immediately said salaam (peace be with you) to each of them. A friendly atmosphere created, I expressed a desire to see the venerable Narbat Singh who had so valiantly journeyed many miles to cast his vote for Chowkooree.

Respectfully lowering their voices they said: "Narbat Singh is very ill; he is even now on his deathbed."

There was a long silence which I did not interrupt. The old men nodded, the thin smoke rose in the clear air, a child called. Then, as though remembering deeds long-forgotten, they told in their reedy voices the story that Narbat Singh had related to them; of his lone journey to Bageshwar and of the brave gentleman who had undertaken to shoot the leopard. But that journey of over thirty miles by footpath and then back to Chowkooree, including many steep and dangerous ascents, had proved too much for his tired old heart. On his return he had complained of feeling exhausted, and had taken to his bed.

Typically hospitable they inquired if I would like some goat's milk. Thanking them, I declined, explaining that I had lunched only a short time ago. Could they bring me something to eat? I replied that my appetite had already been appeased. Then I asked if I might sit with them under the medlars while they gave me the full story of the man-eating leopard.

I began a brief catechism:

"How many people has it killed?"

"We can't say, the Patwari (a petty official) knows."

"Where is he?"

"He has gone to the district headquarters."

"How many persons has thc leopard killed in your village?"

"Three; one of these was Narbat Singh's nephew. He was attacked and taken away when he went to collect fodder."

I asked them to show me the place where this tragedy had occurred, and they pointed westwards. They went on to say that the leopard had killed several of their goats, the last one only two nights ago.

"Where do you think the man-eater wanders about most?"

"He roams in the scrub jungle over there" (pointing west). "When he kills he eats as much as he can and then takes a portion to his cave. He is a lame leopard, so does not go far away, but he is exceedingly cunning."

The mention of a cave immediately roused my curiosity and I asked where it was.

"You can't see it from here," they replied. "Come with us and we will point it out to you."

Rising from the shade of the medlars, the speaker led the way and I followed; behind me in single file were his three companions. Hill folk are ever ready to walk, whatever their age.

We kept to a well-beaten path that brought us to a spring. This was the village's water supply. Round about it the soil was made slushy by the daily scrubbing of brass pots and pans, filling of vessels with drinking water, and washing of clothes. The surplus was diverted to a scooped-out basin lined with clay which served as a trough for cattle.

Beyond the spring was a wide gentle spur on which were fallow fields covered with weeds. Further on there was a narrow ravine shut in by the crowns of moisture-loving trees, giving it the appearance of a broad green ribbon running down the hillside, and fringed by bushes heavily browsed by goats.

Following the leader we came to steepening terrain above and below the path. Halting and facing uphill he pointed to a massive rock jutting out aggressively from the slope of the hill. Around it was a conglomeration of rocks of various sizes. Then, to my delight, he said: "Under that very big rock is the cave where the leopard lives."

Looking at it for a few minutes I realized that, since it formed the roof of the leopard's lair, it would be an ideal position from which to shoot. This was excellent progress, and far better than I had expected.

There and then I formed a definite plan of action for enticing the killer out of his den. I thought of the smoking-out method but it did not appeal to me; it had failed when I had used it on two previous occasions elsewhere, the smoke spreading out so much that it was impossible to aim accurately at the vanishing leopards.

I had reason to be pleased that my quarry had killed a goat so recently, and would thus have sufficient food to make wandering in search of other prey quite unnecessary for at least four or five days. Therefore when my plan went into action the man-eater would almost certainly be at home.

On the whole I had seen enough during the afternoon to make the situation appear extremely promising, and as it was now half-past three it was time to return. Before leaving my companions it was necessary to obtain their co-operation.

I outlined my plan to them, and it met with their full approval. They readily agreed to provide me with a goat the following afternoon at one o'clock, a goat which would bleat loudly and frequently. I added in fairness: "If the goat is killed by the leopard I will gladly pay whatever price you fix." This arrangement impressed them favourably; they said they would help me in every possible way, and it was clearly in their own interests to do so.

They had come a considerable part of the way to my camp

and would have come further, but I advised them to return before darkness set in, since the days were now short. With smiles and salaams we parted to meet on the morrow; they intended going straight to Narbat Singh's house to bring him what comfort they could.

A little later as I sat at my table in my tent enjoying a cup of tea with slices of warm toast and butter, and followed by a refreshingly hot bath, my mind was much occupied with wondering what the next day might have in store, and whether I would have the good luck to get at least a glimpse of this cave-dwelling man-eater.

About a hundred and fifty yards from my camp was a halting place on the trade route, which was habitually used by the Bhotia traders. These men transported merchandise on the backs of sheep and goats and each of them possessed three or four scores of these little animals. When they rested for the night they made pens barricaded with branches and brushwood; they kept a number of savage-looking dogs which had their shaggy necks and throats covered with spiked leather collars to afford some protection from marauders. These vigilant brutes did their duty exceedingly well, often giving their lives in bitter and bloody conflict to save the flocks.

At about half-past eleven that night there was a hue and cry at the Bhotias' encampment. The dogs were snarling and barking, the men shouting, the sheep and goats bleating. I made haste to get a sight of what was happening. By the dull glow of a dying fire it was just possible to discern a leopard doing its utmost to drag a sheep through the barricade; but the general hullabaloo was too much even for this bold intruder, and he abandoned his quarry and slunk away into the surrounding darkness.

Two of the dogs which had fought so bravely were suffering from terrible wounds and were obviously dying, although it was out of the question to attempt to do anything for them. That the killer, however, had not departed unscathed was evidenced by the blood on the iron spikes of their collars.

To prevent the leopard returning, the Bhotias added wood to the fires and planned to remain awake for the rest of the night. I returned to my bed pondering on this turn of events. This was surely not the lame man-eater from Chowkooree? It would not be outside his territory, but he had food for days safely inside his mountain-cave, and would be unlikely to brave the Bhotian dogs unless he really needed food.

I decided to begin work in the morning earlier than usual, and turned the hand of the alarm clock back an hour. I began my inspections at seven, and completed them at half-past eleven. After an early lunch I set out once again for Chowkooree village. No journey then seemed too much for me; I was in fine fettle, traversing the hills for many hours a day, year after year.

On arrival I was glad to see my companions of yesterday waiting for me. With them was a black and white goat, short-horned and displaying a lively manner. In answer to my inquiry they gave me their assurance that it would bleat loudly and repeatedly.

I lost no time in putting my plan into operation. Ascending the section of the hill where the cave was, but taking care to keep away from it in order not to disturb the occupant, if he happened to be there, I went higher and still higher. This time I was leading, then followed the agile goat led by the leanest of the old men, and then his three companions. At this stage I was not displeased that they had apparently decided to exclude any others.

Selecting a salient point, I turned sharply to the right, followed the contour and stopped. Here we were overlooking the rock under which I hoped was a very important resident of the locality. We were about two hundred feet above it.

I informed my followers that I would take with me only the man with the goat; the remainder could relax where they stood and await a signal after I had fired at the leopard.

Accordingly, with the man and the still lively goat I descended slow and as silently as possible until we stood within a short distance of the rock. Here I went forward very cautiously to look for a bush suitably situated about twenty yards from the mouth of the cave. Deciding on one to the right front I signalled to the man to follow me, which he did with some difficulty. I signed to him to tether the goat to the bush. While he was doing this I faced the cave, in case the man-eater emerged.

Having tethered the goat without incident, the man followed me in a detour to the top of the rock, which I found to be wide and flat. I signed to him to sit behind me and there we waited. Not a sound, not a movement. The goat chewed the cud.

After waiting about twenty-five minutes I half turned to my companion, put a hand to my mouth, and pointed to the silent goat. Equally perplexed, he signed that he did not know and shrugged his shoulders in despair.

I sat on that rock for an hour hoping the goat would bleat, but it never did. My companion and I could not conceal ourselves completely from it, and as long as it could see any part of us it knew it was not alone, and so sat down looking trustfully in our direction. A well-fed cat could not have looked more contented.

An attempt at complete concealment showed that the man-eater could have come out of the cave and advanced on the goat without my getting even a glimpse of him. It was indeed a futile situation, not, perhaps, without some little humour in it; two men sitting on a rock above a cave, a goat tied to a bush in front of it and a much-wanted man-eating leopard probably stretched out full length inside the cave with not the least intention of emerging from his lair. Possibly he was enjoying a sound siesta; he might even be enjoying the joke.

But after an hour I had exhausted all the humour in the situation, so untethering the dumb goat, we went, with the same vigilance, back to the waiting men. I explained to them, in full, the cause of my disappointment. They promptly offered another goat, a guaranteed bleater, for the following day, but I told them it was useless while the goat could see us on the rock. I was prepared, as always in these expeditions, to exercise almost unlimited patience, but I owned myself no match for a cantankerous goat. One such failure was sufficient, and in any case, I had thought of a better idea.

On the way back to the village I wondered what they would say when I unfolded my new plan. The sun was nowhere near the horizon; I could spare time in the village to explain my scheme. The men offered me a variety of furniture to sit on; first a stool, then a chair and next a bed. I declined them all very politely for an obvious entomological reason, preferring to sit on the clean slate floor which was swept daily.

Making myself as comfortable as I could, I listened quietly while they told me about Narbat Singh's condition; since the leopard-killer had come he had been staving off his last hour with characteristic tenacity, as though determined not to go first.

Then I could no longer delay in telling them my plan, and a little apprehensively, I asked to see all the boys from ten to fourteen years of age. Intrigued, but polite, they asked for what purpose. I said that I wished to choose one who could best imitate a goat's bleat. There was instant uproar. The men were horrified, and protested vociferously. The women, peeping from doors and windows, shrieked: "Kabhi nahin, kabhi nahin" (never, never). Every father, every mother, every uncle and aunt, at once imagined their son or nephew being forcibly tied to a bush outside the man-eater's cave, made to bleat like a goat, and then being promptly devoured for his pains.

When the noise subsided a little I said: "Now listen to me. This is how it will be. This boy—this fortunate and valiant lad—will sit on the rock above the cave. I will sit on one side of him and his father will sit on the other. I give you my assurance that no harm will befall him. Now call up the boys and let me choose one."

With a better understanding of my intentions, calmer thoughts prevailed but I could still see peeping eyes round the doors and windows, their owners straining their ears to catch every word spoken by me and mumbling very audibly amongst themselves; I asked the men to stop this uncalled-for behaviour. "Chup!" (silence) shouted one of the husbands and instantly there was quiet.

I repeated: "Call up the boys", and the men went off to round them up. Soon they stood in line before me and I looked over the ragged rank and wondered how appealing would be their bleats. As I expected, the boys thought it great fun, and were bursting with giggles. Gone was their apprehension at the sight of a white face.

Beginning from the left I asked each to bleat in turn. At each performance there were peals of laughter—a very different scene from that of a few moments ago. Soon the fathers and uncles were laughing unrestrainedly, and even the women in the shady doorways smiled, a little shamefaced at their previous panic.

I was delighted to witness this good humour and complete change of mood. For me the matter was a serious part of the man-eater hunt; a really good selection could ensure the success of the whole exploit. The novel competition went its way down the file, two or three of them laughing so much that when it came to their turn they could not utter a sound in spite of repeated and much-applauded efforts.

Among the bleating boys were three well-practised ones; they had evidently herded goats and had amused themselves imitating them. From these three I decided to make my choice. They bleated again at my request. I shut my eyes and tried to listen like a hungry leopard. Then I shot out my arm at the most persistent and pathetic and found that I had chosen Jasodh Singh, a lad of fourteen. I hoped his age would help him endure the ordeal before him. He had a bright and merry eye, and was older than his companions—I fancied I saw a likeness to Narbat Singh but refrained from mentioning it in this happy scene.

The senior orderly I had with me, who had been with me on many man-eater hunts, now exclaimed in loud tones "Larka ko kuchh nahin hoga" (no harm will come to the boy). Jasodh Singh shook his head excitedly from side to side, obviously looking forward to the loud bang of a gun. His father, who was also to come, was somewhat reluctant, but more on the boy's account than his own. I was very pleased that the first and vital part of my second plan had made such a satisfactory start, and I returned happily to my camp.

Before leaving the village I asked the men to be ready for me the next day at one o'clock punctually, hoping fervently that no influence, paternal or maternal, would cause Jasodh Singh to change his mind and spoil my second attempt to catch a sight of the leopard.

Sitting in solitude in my tent, sipping refreshing tea and gazing at the heavy official post that had just arrived by runner and had to be disposed of in the next five hours, I found it difficult not to chuckle as that protesting "kabhi nahin, kabhi nahin" repeated itself in my mind. There followed the picture of Jasodh Singh tied to a bush, bleating his very best, the man-eater walking out of his den and seeing a new but welcome kind of goat. . . . Not without sympathy, I thought how vividly this heart-rending scene must have filled the imagination of those many minds when I began to explain my plan and called for a boy who could bleat like a goat!

With sunrise I repeated the time-table of the day before. Completing inspections and quickly lunching, I started out with two orderlies for the village. I was still doubtful about the intentions of the boy's parents. Anticipating the need for a reconversion I was depending on the orderlies to talk them over. A last-minute change of mind was always possible, and for this I had to be prepared.

On arrival I was tremendously heartened by the complete absence of vacillation. It was clear that father and son would abide by their promise and that the whole village was ready to co-operate. The father stood by, awaiting my instructions, and then just as we moved off, a female voice, loud, high and frantic in its intonation, shot out of a window: "You *must* go, he is your son." I had no difficulty in guessing whose voice it was.

We followed the way we took before, Jasodh Singh keeping pace with me. There was no path, and we just walked straight up the hill to the same point above the large overhanging rock. For what I hoped would be the final scene I selected an orderly, the boy's father and a villager to accompany me. I told the other three to wait and watch for a signal. We approached the rock as silently as we could, the father holding his son's hand.

I arranged each one's position on the rock by a series of signs. I sat forward, behind me was Jasodh Singh, then his father, next the orderly and after him the villager. When I was quite ready with my rifle I turned to the boy and, by a shake of the head, asked him to bleat. He obeyed instantly. The imitation was made almost perfect by the faint trill that entered into it, due to a mixture of excitement and nervousness; the bleat, however, was too short; I wanted a longer one. By a wave of the hand I showed him how to prolong it. At once and with the greatest gusto he produced a bleat of incredible length and perfect timbre. It was enough to make any leopard want to know more. But still there was no result and I was beginning to foresee another disappointment.

Then a further idea occurred to me. The sound waves were going into the air and were there being dissipated; whereas if the man-eater was to hear the bleats, they must as far as possible be directed into the cave. To convey this idea to the orderly and Jasodh Singh's father it was necessary to perform almost incredible feats of pantomime, but eventually I made them understand that it was necessary for the boy to lie flat on his belly by my side, with his head over the ledge of the rock, and then to bleat and go on bleating with his face turned downwards and towards the cave's mouth.

Getting into an ideal position to carry out my instructions, with his ankles in the vice-like grip of his two anxious guardians, the intrepid Jasodh Singh spread out his arms and made his position doubly secure by pressing his hands against the rough and severely weathered surface of the rock. He then shuffled slightly forward to enable him to look almost into the cave from a height of about nine feet. I was fairly confident that in theory the sound waves should now reach their target if only he could overcome the handicap of his position. I need have had no anxiety. Looking up at me for a nod he bleated, excelling all previous efforts in volume, in quality, and in repetition. It was a superb performance.

My instructions were that as soon as he saw a head emerging from the cave he was to stop bleating and shuffle silently backwards so that his head would not be seen by the man-eater. He took another breath and the sound wavered over the cave, then quickly died away as he energetically moved about a foot backwards, aided by his two human life-lines.

This was the moment I had longed for. Seconds later out walked the leopard. The limp was unmistakable; this was the man-eater who had brought death to the hillside and unhappiness to my friend Narbat Singh. He stopped and faced downhill, turning his head to the right and left, scanning the bushes for the meal which had come so obligingly close to his lair. There he stood about six yards from the cave, the end of his long tail curved slightly upwards in the form of a wide U and the tip moving very gently from side to side; he was a picture of expectation and hope, but for me a very full and long target, the whole length of his body.

Subconsciously I was aware of Jasodh Singh beside me, but none of us seemed to breathe. Carefully I chose a spot between the powerful shoulders, and then, downwards and obliquely, the bullet crashed into him, just where it was intended to go. Instantly he slumped forward, his body sprawled downhill and rolled and rolled till it came to rest against a rock many feet below on the long slope, and there lay quite still.

My first thought, after lowering the rifle and extracting the empty cartridge, was for Jasodh Singh. How courageously, at the age of fourteen, he had acted his role! Feeling a tremble in his body I quickly drew him to me, made him sit up and kept him close against me until the trembling ceased and he was quite himself. The sight of the leopard followed by the loud report of the rifle had been a little too much for him.

I pressed a handsome reward into his hand and another into his father's. I have always felt that a reward given on the spot is worth double a delayed one and is certainly far more appreciated.

Standing up on the accommodating rock I signalled to the waiting men to come and join us. We walked down, pulled the man-eater away from the rock and had a good look at him. The cause of his lameness was difficult to diagnose; it was probably due to a permanent dislocation of a joint since there were no outward signs of injury. Otherwise there was nothing wrong with him. His forepaws were intact, so were his teeth, even his incisors were complete in number. He measured seven feet four inches and he had a good coat. I concluded that only his lameness had driven the leopard to man-eating.

I made a very cautious inspection of the cave which had sheltered him so well. It was deep and wide so that no rain, however slanting, could penetrate far, and it was quite dry. It was an ideal lair, and one which no doubt would soon harbour another of his kind, unless forestalled by a bear for whom it was equally well suited. The stench was overpowering. There were bones of divers descriptions scattered about and part of a human foot in an advanced stage of decay; there were also shreds of cloth lying about.

What we might have found in the inner part of the cave I do not know for the stench was such that it effectually damped our curiosity, and forced us to retreat.

The men carried the leopard down in a triumphal procession to the village, and deposited it on the floor of the front courtyard for all to see. The women and children, who had all disappeared when they heard us approaching, appeared cautiously from their seclusion; but I noticed an equal reluctance among the men, who stood about in a silent knot beneath the medlars. I knew in my bones what had happened, and walked over to them. Narbat Singh, the tired old man, had quietly slipped away from life while we hunted his enemy on the hillside. I felt pleased that I had vindicated the vote he had given me.

Leading the village men over to the man-eater, I saw that Jasodh Singh's mother had her arm round her son and was laughing softly with her husband.

Calling the boy to my side and placing a hand on his head, I hailed him as the hero of the day, and told the whole gathering that through his valiant behaviour the man-eater would prowl no more in their village. I felt sure that Narbat Singh's atma (spirit) was present with us all to share our rejoicing.